I0015390

The 3CX IP PBX Tutorial

Develop a fully functional, low cost, professional PBX phone system using 3CX

Matthew M. Landis

Robert A. Lloyd

PUBLISHING

BIRMINGHAM - MUMBAI

The 3CX IP PBX Tutorial

Copyright © 2010 Packt Publishing

All rights reserved. No part of this book may be reproduced, stored in a retrieval system, or transmitted in any form or by any means, without the prior written permission of the publisher, except in the case of brief quotations embedded in critical articles or reviews.

Every effort has been made in the preparation of this book to ensure the accuracy of the information presented. However, the information contained in this book is sold without warranty, either express or implied. Neither the authors, nor Packt Publishing, and its dealers and distributors will be held liable for any damages caused or alleged to be caused directly or indirectly by this book.

Packt Publishing has endeavored to provide trademark information about all of the companies and products mentioned in this book by the appropriate use of capitals. However, Packt Publishing cannot guarantee the accuracy of this information.

First published: February 2010

Production Reference: 1020210

Published by Packt Publishing Ltd.
32 Lincoln Road
Olton
Birmingham, B27 6PA, UK.

ISBN 978-1-847198-96-9

www.packtpub.com

Cover Image by Vinayak Chittar (vinayak.chittar@gmail.com)

Credits

Authors
Matthew M. Landis

Robert A. Lloyd

Reviewer
William England

Senior Acquisition Editor
James Lumsden

Development Editor
Rakesh Shejwal

Technical Editor
Mazhar Shaikh

Ishita Dhabalia

Indexer
Rekha Nair

Editorial Team Leader
Gagandeep Singh

Project Team Leader
Priya Mukherji

Proofreader
Jade Schuler

Graphics
Geetanjali Sawant

Production Coordinator
Shantanu Zagade

Cover Work
Shantanu Zagade

About the Authors

In 1995 **Matt Landis** started Landis Computer, which has been providing IT services to small businesses for 14 years and is now a 11 person Microsoft Gold Certified Partner. Matt has over 14 years of field experience implementing Windows Server, Microsoft & Dynamics ERP solutions in small business environments. Landis Computer was the first company in the USA to be designated a 3CX Premium Partner.

Matt is very active in the Windows-based IP PBX community: he is both a 3CX Valued Professional and pbxnsip Certified, has contributed thousands of posts to the 3CX community forum, hosts the http://windowspbx.blogspot.com blog, and writes a monthly Windows IP PBX e-newsletter for VARS.

Matt also has various general IT certifications: Microsoft Certified Systems Engineer, Microsoft Certified Database Administrator, Microsoft Office Certified Expert, Microsoft Certified Dynamics, and Network+ and A+.

When not working and when he can afford a chance Matt likes to travel internationally with his wife Rosalyn and is very involved with his church.

First, I would like to thank God for the capability & opportunity to do interesting things like write this book. And secondly, I would like to thank my wife Rosalyn for being very supportive in spite of all the time it took.

I would also like to say thanks to Rob for co-authoring—it was fun! And of course, I would like to thank the whole 3CX community for all the time they are willing to share helping others.

Robert Lloyd has been in the IT field for 20 years. He graduated college with a B.S. degree in Computer Science, and holds many certifications including MCSE 2003: Security, MCTS on Server 2008, Vista, and Exchange 2007, Small Business Specialist, A+, Security+, and Cisco CCNA. He has been running his own consulting business, TechNet Computing, for five years. Before that he worked for a large law firm as the IT Director for almost eight years, and also developed computer based training software.

Rob also teaches Microsoft, Cisco, and CompTIA certification classes and is a Microsoft Certified Trainer (MCT). For the past five years he has taught at Today's Tec as the lead instructor.

Rob has been involved in VoIP for four years and has been using 3CX since version 3. Rob has contributed to helping others install, configure, and troubleshoot their own systems online and remotely.

I would like to thank my family and friends for their support and inspiration in this project. I would also like to thank Matt, without him this project would not have been done. It was fun working with you Matt! I'd also like to thank 3CX. They have pioneered VoIP for the Windows community.

About the Reviewer

William England (WJE) started his professional IT background in 1991 after receiving training from a leading software house in Unix and Xenix before moving on to Microsoft Operating systems with the advent of Windows NT and further releases of Windows.

In 1995, William set up the IT and networking arm of William J England & Son Ltd, offering various IT-based services from installation, support and consultancy for local and overseas companies.

With the increasing need for e-mail and enhanced telecommunications William extended his area of expertise to GSM cellular distribution to complement the existing service level, together with e-mail and hosting solutions, anti spam, content security, and eventually VoIP.

Table of Contents

Preface

3CX was one of the first Windows-based PBX phone systems in the market. Even today, there are only a couple out there that work well. Traditional PBX phone systems are "black boxes" that get mounted to a wall, and you can't do anything with them. If you want an upgrade, you call the vendor. If you want more features, you call the vendor. If you want to make a change, you either call the vendor or you learn how to use their command-line system from a 1000-page manual. Either way, it's expensive and time consuming. 3CX was created to make phones easy!

3CX will run on any current (XP or higher) Windows platform. It is easy to install, has a terrific GUI, and changes and upgrades are pretty painless. There are, of course, some things to look out for that this book will cover but, overall, it's a great product.

What this book covers

This book is designed to cover everything you need from start to finish and then how to troubleshoot once you're done.

Chapter 1, *Getting Started with the 3CX Phone System*, covers what 3CX is, compares Asterisk with 3CX, and also compares the different versions of 3CX. Then, we get into the components needed and some capabilities.

Chapter 2, *Downloading and Installing 3CX*, will teach you what are the requirements to get 3CX to work well, the hardware requirements, and some points about operating systems (Windows). We will also cover downloading and installation and some of the options available. Then, we cover one of the most important parts — how to log in to the interface.

Chapter 3, *Working with Extensions*, covers the different types of phones you can integrate with 3CX. Also, we dive into creating extensions and just about all the features you can configure.

Chapter 4, *Call Control: Ring Groups, Auto-attendants, and Call Queues*, teaches you how to configure 3CX to handle calls once you have your extensions set up. We cover Digital Receptionists, Dial by Name directories, Call Queues, and Hunt groups.

Chapter 5, *Trunks: Connecting to the Outside World*, covers SIP trunks, PSTN lines, and some features to look for while selecting and using both of them.

Chapter 6, *Configuration*, covers several of the many features available in 3CX, from creating Music on Hold to prompt sets and dial plans. We also cover DIDs for those extensions that want the call going right to their phone.

Chapter 7, *Enterprise Features*, covers how to set up and use the enterprise features that come with the paid license of 3CX. Features like call recording, conference calls, call reporting, and faxing will be discussed. We will also cover the mystery behind codecs and how/when to change them.

Chapter 8, *Integrating 3CX*, covers the various types of integration options available with 3CX. We will cover topics like Exchange 2007, Skype, instant messaging, dialing from Outlook, and database integrations.

Chapter 9, *Hardware*, is an important chapter if you are looking to buy hardware. While we discussed hardware phones before, this chapter breaks them down into a few brands that we know and use. We will also cover devices to connect to your analog phone lines and to analog phones.

Chapter 10, *Maintenance and Troubleshooting*, covers some troubleshooting options that you will need at some point, once your system is up and running (or maybe not). We will also go over disaster recovery and backing up your system. Then, we will move on to deeper networking with firewalls, network services, logging, and finally support options when you are really stuck.

What you need for this book

We are assuming that you have some Windows experience, access to your router, and that you know how to make changes to your firewall (if needed). You will also need a DHCP server, some kind of VoIP phone (hardware or software), and high-speed Internet or old school phone lines. We will cover all of these throughout this book, but the more you know to start, the easier it will be for you to get your system up and running quickly and easily.

Who this book is for

Anyone familiar with Windows and some basic networking knowledge can install and set up a complete 3CX Phone System. This book takes you through all the steps you'll need, plus some tips and tricks to make it better. We'll also cover some topics to make it easier using third-party applications.

If you typically never open a manual like the one 3CX has, this book will help you. The knowledge Matt and Rob put into the book are based on real-world examples.

If you have tried another phone system, such as Asterisk, but had issues with commands or integration, then 3CX and this book are for you. If you follow all the chapters, you will have a fully functional system in a short time.

This book is not for those who hate Windows or who know nothing about networking. If you don't know what a TCP/IP address is, you can still get a functional system, but if you run into problems, it may be hard to troubleshoot.

Conventions

In this book, you will find a number of styles of text that distinguish between different kinds of information. Here are some examples of these styles and an explanation of their meaning.

Code words in text are shown as follows: "3CX VoIP Client will run the program `c:\getgpdata.exe` with `%callid%` as the parameter."

New terms and **important words** are shown in bold. Words that you see on the screen, in menus or dialog boxes for example, appear in the text like this: "The following screenshot shows the **Preferences** interface, which is used to set up the powerful **On Incoming call** feature."

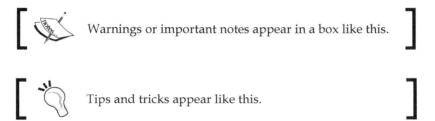

Warnings or important notes appear in a box like this.

Tips and tricks appear like this.

Reader feedback

Feedback from our readers is always welcome. Let us know what you think about this book—what you liked or may have disliked. Reader feedback is important for us to develop titles that you really get the most out of.

To send us general feedback, simply send an e-mail to feedback@packtpub.com, and mention the book title via the subject of your message.

If there is a book that you need and would like to see us publish, please send us a note in the **SUGGEST A TITLE** form at www.packtpub.com or e-mail suggest@packtpub.com.

If there is a topic that you have expertise in and you are interested in either writing or contributing to a book on, see our author guide at www.packtpub.com/authors.

Customer support

Now that you are the proud owner of a Packt book, we have a number of things to help you to get the most from your purchase.

Errata

Although we have taken every care to ensure the accuracy of our content, mistakes do happen. If you find a mistake in one of our books—maybe a mistake in the text or the code—we would be grateful if you would report this to us. By doing so, you can save other readers from frustration and help us to improve subsequent versions of this book. If you find any errata, please report them by visiting http://www.packtpub.com/support, selecting your book, clicking on the **let us know** link, and entering the details of your errata. Once your errata are verified, your submission will be accepted and the errata will be uploaded on our website, or added to any list of existing errata, under the Errata section of that title. Any existing errata can be viewed by selecting your title from http://www.packtpub.com/support.

Piracy

Piracy of copyright material on the Internet is an ongoing problem across all media. At Packt, we take the protection of our copyright and licenses very seriously. If you come across any illegal copies of our works, in any form, on the Internet, please provide us with the location address or website name immediately so that we can pursue a remedy.

Please contact us at copyright@packtpub.com with a link to the suspected pirated material.

We appreciate your help in protecting our authors, and our ability to bring you valuable content.

Questions

You can contact us at questions@packtpub.com if you are having a problem with any aspect of the book, and we will do our best to address it.

1
Getting Started with the 3CX Phone System

Welcome to this book on the 3CX IP PBX software-based phone system. In this chapter, we'll take a look at what 3CX is and where 3CX fits in the big picture of modern phone systems. We'll do a couple of comparisons such as hardware-based phone systems versus software-based phone systems, Asterisk compared to Windows-based systems, and 3CX Free edition compared to 3CX Commercial edition. We will take a look at the major components of the 3CX Phone System and at what is not included in 3CX. We'll also discuss a bit about the company—3CX.

In this chapter, we will take a look at the following:

- About the 3CX company
- What the 3CX Phone System is
- Comparing hardware-based and software-based phone systems
- Asterisk versus Windows Phone System
- Major components of the 3CX Phone System
- Some characteristics of the 3CX Phone System
- What the 3CX Phone System is not

About the company—3CX

3CX is relatively new to the telephony world being founded in 2005 by Nick Galea as compared to Asterisk in 1999, pbxnsip in 2001, and Objectworld's UC Server product in 2001. Nick is a seasoned software entrepreneur having started several other well-established companies including 2X (thin client software) and Acunetix (web security).

What does the name 3CX stand for? According to 3CX CEO, Nick Galea, 3CX stands for **Connect**, **Communicate**, and **Collaborate**.

What the 3CX Phone System is

Simply put, the **3CX IP (Internet Protocol) PBX (Private Branch Exchange)** is a phone system for an organization or business. Traditionally, phone systems were a proprietary piece of hardware, designed just to do the functions of a phone system. 3CX is the software that can turn any standard Windows personal computer or server hardware into a complete phone system.

The 3CX Phone System does all the things that normal phone systems do. It allows phones connected to it to call each other and call external phones, too, whether they are on the good old **Public Switched Telephone Network (PSTN)** or a **Voice over IP (VoIP)** network. Wow, that sounds simply fantastic, but what a phone system is really designed to do is allow two or more people to talk to each other.

Phone systems have many features to assist in the effort of helping people talk on the phone (or maybe I should say *communicate* in the case of voicemail) and 3CX is no exception. Calls can be transferred, put on hold (with music playing), sent to a menu of options that callers can select from, sent to a queue to wait until a customer service representative can help them, or sent to voicemail. Calls can be routed to a digital receptionist, an extension, voicemail, or even to an external mobile phone based on a schedule of times.

I mentioned earlier that 3CX is a VoIP phone system. This means that your voice traffic goes over the same Ethernet cables that your computer network traffic does. As phones' handsets or softphones are connected to the 3CX Phone System via Ethernet, this means you can have a remote phone at your house connected to your office phone system!

 A very common misconception related to VoIP phone systems is the belief that, if you use one, you also need to use VoIP phone lines to connect to the outside world. It is very common for a company to use VoIP internally to avoid wiring their office with both Ethernet and RJ11 telephone lines and use a regular telephone line for external calls because they may not have a good, fast broadband connection.

3CX supports multiple gateway devices that translate PSTN to **Session Initiation Protocol (SIP)**, so that you have total freedom when deciding what you want to use. Do you have plenty of broadband speed? Then save some money and go with a VoIP phone line. If not, stick with your good old PSTN telephone line. The fact that 3CX can use PSTN and VoIP is what we want to underscore.

One of the strong points of 3CX is its ease of use, which when talking to 3CX users, is the number one reason people use it. With 3CX, setting up a phone system can be so easy that even a non-telephone user can do it.

While there is a free edition of 3CX, it is not open source or GPL, but rather a commercial product. The free edition of 3CX is not time limited but does have some feature limitations. There are hobbyists and small businesses that use the free edition of 3CX in their day-to-day business or home phone system.

As we are on the subject of free edition, I thought it would be better to mention that 3CX can be used to create an entirely free phone system. If you use the free version of the 3CX Phone System, then by using a 3CX softphone installed on your computer and a SIP VoIP phone line provider, you can build an entirely software-based and free phone system. (In full disclosure, I should mention that you will probably want to get a good headset—so maybe not quite free!)

Another defining part of 3CX is its openness and ability to work with hardware, such as phone handsets and gateways from many vendors. 3CX uses an industry standard protocol called SIP to talk to devices. So, any device that uses SIP standards can theoretically work with 3CX. A list of supported devices to choose from are available at `http://wiki.3cx.com/phone-configuration`.

A conversation about 3CX is not complete without a comment about the 3CX forum community as a resource that adds a lot of value to 3CX. While 3CX is not open source, the forum is very active and helpful with thousands of excellent and helpful posts. Don't overlook this valuable part of your 3CX system: `http://www.3cx.com/forums/`.

3CX has also developed video training for the 3CX Phone System, which can be found at `http://training.3cx.com`.

Hardware versus software phone systems

Traditionally, phone systems were a piece of hardware designed to be a phone from day one. Also the phone handsets and everything else involved with the phone system was designed by the same vendor to work with the telephone box. Most often you couldn't take a handset from one phone system vendor and plug it into a phone system from another vendor. In other words, the vendors were using proprietary protocols and communication methods. One nice aspect about this "one vendor" design is that, when you got a bundle, it was made to work together and you had (at least theoretically) few interoperability issues. This, of course, came with a price in dollars and limited the ability to integrate the phone system with the rest of the computerized things going on in your office.

Then Asterisk came along and changed people's expectations about what phone systems should cost and what they should be able to do in terms of being integrated with existing computer systems. Now, instead of paying thousands of dollars for a phone system, you could take a free version of Asterisk and load it on an aging server that you took out of service and have a phone system at a low cost. At first, only experienced technical people used Asterisk because there was a lot of command line and editing text files involved, but then easier-to-use web interfaces were added to remove some of the complexity.

The following table compares hardware-based and software-based phone systems:

Hardware-based phone systems	Software-based phone systems
Complete bundle	Not a bundle — integration of components is required
More Costly	Less cost up front
Not well-integrated with your computer system	Integrates well with your computer system
Usually installed by a specialist	Because of familiarity with Windows, it is more common for hobbyist or businesses to "do it yourself"
Rock solid day-to-day operation (99.99% uptime?)	Less rock solid day-to-day operation
Support contract for ongoing maintenance	Often supported by the company IT person

Linux Asterisk versus Windows 3CX

Inevitably, when talking about free, software-based phone systems, the extremely popular Linux-based Asterisk IP PBX will come up as an option. Should we use Asterisk or a Windows-based solution? Let's look at some of the strong points of each.

The following are the advantages of Asterisk IP PBX:

- It has been around for a while and has a lot of installs
- There are a lot of add-on products surrounding it
- Asterisk installs on a free operating system—Linux
- Asterisk Free version has features comparable to 3CX Commercial edition

The following are the advantages of Windows 3CX:

- Large number of people already familiar with Windows server administration
- Easier to integrate with existing Windows networks
- Has a very consistent and easy-to-use user interface
- Easier to integrate with Microsoft Exchange
- Supports SIP forking (allows two or more devices to be registered to the same extension)

I started out in the IP PBX world using **Asterisk@Home** (now **Trixbox**). Because of my unfamiliarity with Linux, I spent a lot of time trying to figure out how to perform simple operating system tasks. After working with Asterisk for quite a while, it dawned on me that what I really wanted was a free (or low cost) phone system with integration capabilities, one that worked reliably, and one that I could call for support in rare instances when something didn't work right.

 At one point, getting commercial support for Asterisk was an issue, but now commercial support options are available. There still may be an issue to get a vendor who will give a list of hardware that they will support along with Asterisk. The "last straw" that made us switch to a Windows-based phone system was when we had an issue with Asterisk. We called several vendors and asked them what hardware should we get to be in a supportable condition, and no vendor could list it for us. However, once again, I believe support for Asterisk is no longer an issue.

I think the decision between 3CX and Asterisk is one that each administrator needs to make. However, for many small businesses, a Windows solution makes sense.

3CX Free versus 3CX Commercial edition

One thing that makes 3CX very attractive to hobbyists, small businesses, and IT managers is its cost of entrance—free! The 3CX community has grown quite rapidly because of the free edition of the 3CX Phone System. It can be a little confusing to a new 3CX administrator about what is included in both the Free and Commercial editions, but the chart provided by 3CX at the following URL helps clarify a lot of questions that come up. The latest feature comparison chart is available at http://www.3cx.com/phone-system/edition-comparison.html.

Major components of the 3CX Phone System

The 3CX IP PBX Phone System is made up of several major components:

- The 3CX Phone System consists of several Windows services, an SQL database to store configuration data, and a web interface for administration. The 3CX VoIP Client and 3CX VoIP Phone are softphones that allow us to use a computer coupled with a headset as a phone.

- The 3CX Call Assistant is a software operator panel that allows us to see the status of the phone system, control phone calls, and do simple chats between the operator and the caller.

- The 3CX Call Reporter allows us to print graphs and reports of call details.

3CX Phone System

After installing 3CX, the first place you'll visit is the administration web interface. This interface (GUI) allows you to set up and maintain your 3CX Phone System. It also helps a 3CX administrator to restart services that get hanged or stuck, even if the services are remote, as shown in the following screenshot:

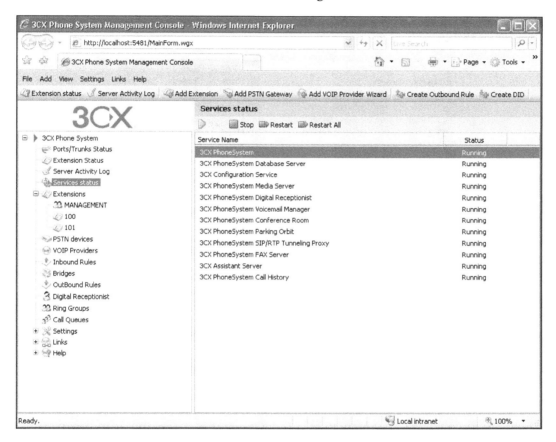

The 3CX Phone System is made up of 13 Windows **Services** that make up the core of 3CX, as shown in the next screenshot. These are **Standard** Windows services, and you can use normal Windows administrator tools to work with them.

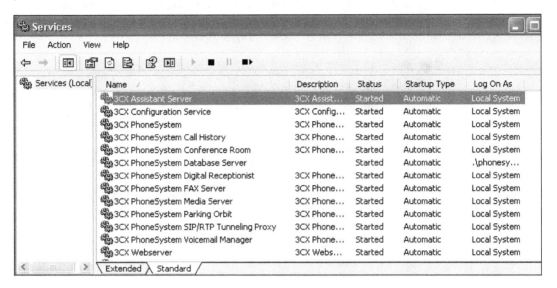

A web system management console provides a simple way to administer and see the status of the system.

A very common question I hear is: "Can I disable this or that 3CX service?" The answer is "No." If you are very smart, you may be able to figure out how to save yourself a few kilobytes of RAM. Most likely, you will spend a lot of time thinking and in the end leave it running. They are all designed to run even if you are not using them.

There are three ways to navigate in 3CX—the navigation pane, the drop-down menus, and the quick launch toolbar.

The navigation pane

The navigation pane has been a part of 3CX navigation since the beginning and allows you to navigate everywhere in 3CX. While some of the other navigation methods allow quick access to adding new extensions, PSTN devices, and other new objects, the navigation pane is the only method to navigate to an existing object and edit it.

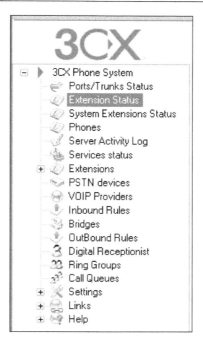

The previous screenshot shows the navigation pane, and the objects in the list are described as follows:

- **Ports/Trunks Status**: See the status of your PSTN and VoIP phone lines/trunks.

- **Extension Status:** See the status of each extension on your system.

- **System Extensions Status**: See the status of 3CX system extensions.

- **Phones**: See and manage your hardware phones.

- **Server Activity Log:** See the status of your server and any errors in almost real time.

- **Services status**: See the status and stop, start, or restart the 3CX Windows services.

- **Extensions**: View, add, or edit your extensions.

- **PSTN devices**: View, add, or edit your PSTN devices.

- **VOIP Providers**: View, add, or edit your VoIP providers.

- **Inbound Rules**: Set up where inbound **Direct Inward Dialing (DID)** is routed.

- **Bridges**: Set up connections between 3CX phone systems.

- **OutBound Rules**: Set up how outbound calls are routed and to which PSTN or VoIP provider.
- **Digital Receptionist**: Set up menus which callers can navigate through.
- **Ring Groups**: Set up groups of extensions that can ring together.
- **Call Queues**: Set up call queues.
- **Settings**: This is where general settings like Music on Hold, office hours, and dial codes are set. This is also where the license is activated if needed.
- **Links**: This provides links to useful features, such as the 3CX softphone download, purchasing 3CX, and more.
- **Help**: This provides links to 3CX's FAQ, forum, blog, and more.

Drop-down menus

Drop-down menus provide an alternate method of navigating to most objects in 3CX. Navigation is divided into logical groups as follows:

- **File**: Allows you to log out
- **Add**: Adds new objects such as **Extension, PSTN Gateway,** and so on
- **View**: Allows you to view the status of your system
- **Settings**: Changes global 3CX settings like network, fax, and system prompts
- **Links**: Provides links to downloads and updates
- **Help**: Provides links to manuals, guides, and support for 3CX

The following screenshot shows us the drop-down menu for the **Add** group in 3CX:

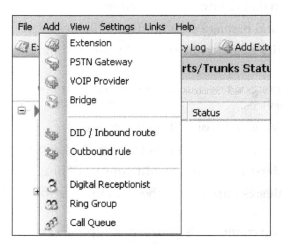

Quick launch toolbar

The quick launch toolbar provides a one-click method to navigate to some of the most used objects in 3CX: **Extension status**, **Server Activity Log**, **Add Extension**, **Add PSTN Gateway**, **Add VOIP Provider Wizard**, **Create Outbound Rule**, and **Create DID**. The following screenshot shows the quick launch toolbar:

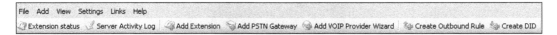

3CX Phone

The 3CX Phone is a SIP softphone that will allow you to use your computer coupled with a headset as a replacement for a desk phone. The 3CX Phone is much like other softphones available and includes similar features. Some of the features included are as follows:

- Take and place calls
- Handle multiple calls (three at a time)
- Place calls on hold
- Transfer calls
- Show incoming calls
- **Telephony Application Programming Interface** (**TAPI**) driver for dialing from Microsoft Outlook (Not free)
- It is a standard SIP softphone, so it works with any SIP-based IP PBX
- The call recording button saves a sound file on the local computer or in the 3CX Phone System
- Auto-answer on paging is supported
- Supports some wireless headset call pickup buttons (such as Plantronics)

- • Unlike the earlier softphone from 3CX (3CX VoIP Client), no presence indication is built-in because it is assumed the 3CX Call Assistant will be used if needed

The 3CX Phone has a very thorough call logging built-in, and **Missed**, **Answered**, **Dialed**, **Recorded**, and **All calls** lists are available. We can see the quantity and duration of the different groups of calls, as shown in the following screenshot:

The 3CX Phone also does call recording at two places—recording calls to the local PC hard drive or to the 3CX Phone System, so that they show up in the 3CX user portal. Looking up **Calls recordings** on the local PC is shown in the following screenshot:

 The 3CX Phone is a standard SIP softphone and can be used with any standard SIP provider or IP PBX. In fact, there is a 3CX forum dedicated to help those who may be using the 3CX Phone in non-3CX Phone System scenarios.

3CX Assistant

In short, the 3CX Assistant gives you, the user, a visual indication of what is happening with your phone system. The 3CX Assistant will also allow us to do some call control by dragging and dropping objects. It is a software version of what, in the good old days, was a hardware device that the receptionist used to direct calls. The 3CX Assistant is a new addition to the 3CX suite. The road map is for the 3CX VoIP Phone and the 3CX Assistant to eventually replace the 3CX VoIP Client. Following is a list of indication features:

- The status of all connected **Extensions**
- The status of all **Queues**
- The status of all **Parked calls**

- Allows visual grouping of extensions
- Voicemail indicator (along with the number of voicemails)

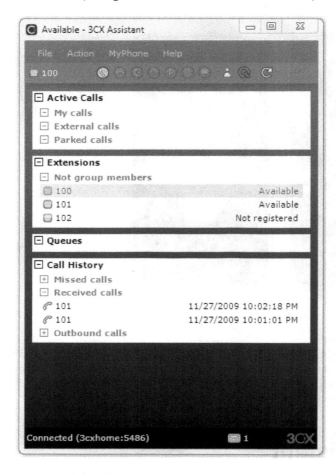

Following are a few of the 3CX call control features:

- Set extension to **Busy** or **Available**
- Log in or log out to/from **Queue**
- Park or pick up a parked call
- Divert an incoming call to voicemail
- Transfer a call by drag and drop
- Record a call
- **Barge in** to a call

Following screenshot shows the **3CX Assistant Action** menu:

The 3CX Assistant also provides a way to integrate incoming calls with your **Customer Relationship Management (CRM)** or some other software package by specifying a program to run and allowing you to pass the caller ID to it as a parameter. The following screenshot shows the **3CX Assistant Configuration** screen:

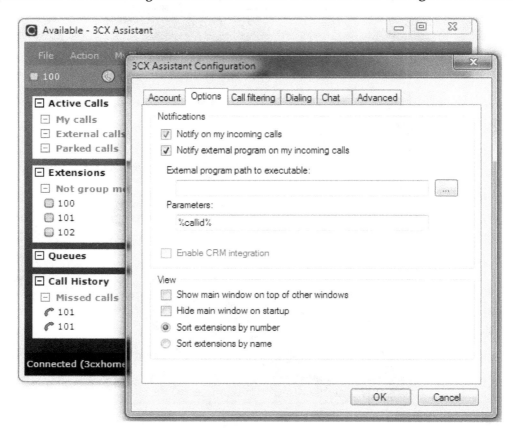

The 3CX Call Assistant also includes a nice and simple chat client, as shown in the next screenshot. This allows users to communicate via instant messaging without installing a full-fledged instant message server like Openfire or Microsoft Office Communications Server.

3CX VoIP Client

The 3CX VoIP Client is a SIP softphone like the 3CX softphone and is an older product that is being sunsetted by 3CX. Because it has some unique features, we will mention it here. The following screenshot shows a **3CX VoIP Client**:

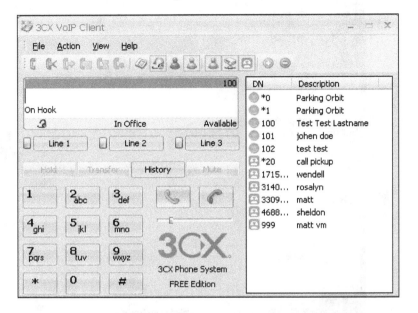

While the 3CX VoIP Client has many features similar to other standards-based SIP softphones, it also has features that are proprietary to 3CX and helps make it more tightly integrated with the 3CX Phone System. Few of the features are as follows:

- Ability to show the status of other extensions (Not free)
- A red light means the other extension is not registered to 3CX
- A green light means the other extension is registered to 3CX
- A yellow light means the other extension is on a call
- A black light means the other extension has been set to **Away**
- Tunneling all VoIP traffic over a single TCP port
- A button to divert an incoming call to voicemail
- A button to toggle **Away** or **Available** status to publish simple presence information to other extensions
- A button to start recording this conversation that will save the recording on the 3CX server as opposed to just saving a sound file on the local computer (Not free)
- Queue status monitoring (Not free)

The following screenshot shows the **Preferences** interface, which is used to set up the powerful **On Incoming call** feature:

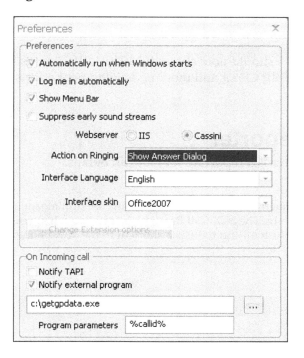

The 3CX VoIP Client has one powerful feature that is not included with most other free SIP softphones. It has the ability to run a program or web page and pass the caller ID of an incoming call to that program or web page. All you need to do is check the **Notify external program** checkbox, enter the path of the program, and add any **Program parameters**. In the previous screenshot, 3CX VoIP Client will run the program c:\getgpdata.exe with %callid% as the parameter. If the phone number calling you was 1-800-555-8383, Windows would run c:\getgpdata.exe 18005558383. You can also use this method to open a web page. Think of the integration possibilities! This is a powerful feature and is available for free.

 Note that 3CX VoIP Client **On Incoming call** feature works with other IP PBXs too.

Because of the lack of wide microphone support in Terminal Services, 3CX does not support running the 3CX softphone or 3CX VoIP Client on Terminal Services.

One last thing that we should note about the 3CX VoIP Client is that 3CX is sunsetting the 3CX VoIP Client and moving development efforts to the newer 3CX VoIP Phone.

3CX Call Reporter

The **3CX Call Reporter** is a reporting tool for call details.

 Note that most of the 3CX components are downloaded and installed separately. The 3CX Call Reporter is installed when you install the phone system, and there is a shortcut in the 3CX Phone System's **Start Menu** folder.

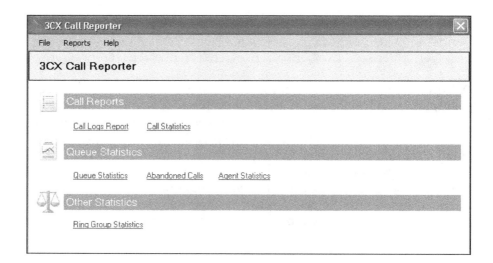

 The Call Reporter is not a real-time tool. There is an update process that needs to be done before the call detail records will be available to the Call Reporter. In version 6, this was real-time, but this was changed to improve performance.

3CX Gateway for Skype

The 3CX Gateway for Skype allows us to use Skype to make calls on our 3CX Phone System. Originally, this was an add-on module to 3CX, but we are including it in this section as it has been integrated with 3CX. You can add a **Skype Trunk Line** just like a PSTN or SIP trunk line as shown in the following screenshot:

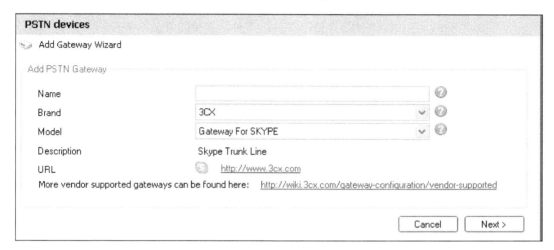

 I don't know what will be the need for this add-on in future, considering Skype just released a SIP to Skype server. We'll wait and see!

3CX Hotel module

The 3CX Hotel module is a web-based application that adds a full **Hotel PBX** to 3CX. It provides features such as wake-up calls, check-in and check-out, guest call log printouts, and room service can set the room availability via a phone call.

 The Hotel module is unique because it is purchased as a separate license from 3CX. All the other modules mentioned are included at no extra cost.

Some characteristics and features of 3CX

There are several major characteristics and features of 3CX that we'll take a quick look at.

Easy to use

3CX's ease of use is probably the number one characteristic that is most beneficial to users. Being Windows based, it immediately resonates with many 3CX users. I find it interesting that a lot of users need to be reminded to look at the manual for more complex tasks because they get so used to doing things without it. One of the most complex tasks in any software-based phone is setting up a gateway device. In many phone systems, this involves logging into the device, configuring a host of settings for the device, and often using trial and error to see which setting works. In 3CX, there is a wizard for most common gateways that will create a configuration file, and you just have to import it into your device. I have never seen a software-based phone system that makes configuring a gateway so straightforward, as you can see in the following screenshot:

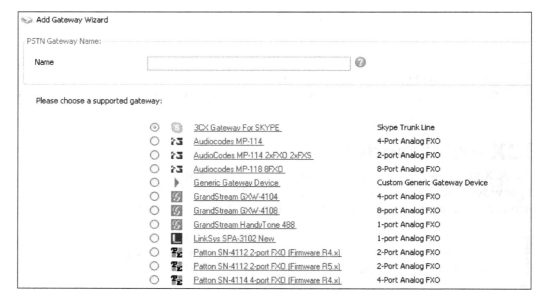

Open and vendor independent

3CX also does an incredible job at documenting how to set up phone handsets and other hardware with the 3CX Phone System. If a device uses SIP standards, it can usually be made to work with 3CX with enough time, but you will save yourself a lot of sweat and tears by sticking to 3CX-supported devices. If you ever get frustrated with the documentation about a phone model, just remember—no other software phone system has as good phone configuration documentation as 3CX.

This question comes up often: "I have such and such SIP phone, will it work with 3CX?" The answer is: "Any standard SIP device should be able to interoperate with 3CX." So, if you have enough time, you will eventually be able to make it work. Did that sound like a warning? It was. My suggestion is to only use hardware supported by 3CX. You can find that list at http://wiki.3cx.com/phone-configuration.

Another question I often hear: "I'm using a supported phone model XYZ. My XYZ feature doesn't work." The answer is: "Not all features on supported phones will work." Take a look at the interoperability notes at http://wiki.3cx.com/phone-configuration/firmwares-tested to help you decide which phone handset will work for you. The phone I personally use to avoid as many problems as possible is **Snom**. Also, make sure you have tested the phone for any firmware.

Windows-based

Another core characteristic of 3CX is that it is Windows-based. There is no Linux or Apple version of 3CX, and this is by design. 3CX (the company) made a very conscious decision to be a Windows product. Initially, and up until version 6, 3CX was based on PostgreSQL and Apache, which seemed somewhat misaligned with the Windows-only policy. In version 7, 3CX showed its increasing commitment to the Microsoft stack and did a major rewrite of 3CX using the Microsoft **Internet Information Services** (**IIS**) web server. Although there have been rumblings in the past, there is no official mention of a move to Microsoft SQL at this moment.

Sometimes this question is asked: "Even though 3CX is very Windows-centric, why doesn't 3CX integrate with Active Directory?" The answer is: "It may, at some point". Right as 3CX stands, it will run on Windows XP Professional just as good as Windows Server, so it can't be dependent on Active Directory.

What the 3CX Phone System is not

Perhaps the best way to understand a product is to take a look at some features the product doesn't have. Those features are discussed next.

3CX is not expensive

Certainly, the free edition is free, but the commercial edition is very reasonably priced, too. What makes the commercial edition quite reasonable is the licensing method of **per concurrent call**. This means that 3CX is not licensed as per extensions attached to the system but by how many concurrent calls (or simultaneous, calls as 3CX calls them) can be made at the same time. For example, if you have a very low-usage phone system (like a retail store) where you might need 50 phone handsets but rarely more than 5 are being used at one time, you could get a 10 concurrent calls user license in 3CX. With competing systems you would need a 50 concurrent calls user license. While this does reduce the cost of 3CX, the issue of what constitutes a concurrent call becomes very important, so make sure you understand it. One factor that is often overlooked is internal calls do count against concurrent calls. To see a full breakdown of what constitutes a concurrent call, I suggest that you read the following post by 3CX engineer Kevin Attard:

```
http://www.3cx.com/forums/what-is-a-simultaneous-call-8123-
15.html#p41947
```

3CX is not a Cisco level of maturity product

Remember the price? So, what does this mean in the day-to-day operation of 3CX? You might need to reboot your 3CX server occasionally, you might need to use a workaround because some feature may still be a work in progress, or some seldom used voice prompt might not have a "million dollar" sound. In short, there will be items you will need to work through. My suggestion is that if you are installing a phone system for a company with toleration only for perfection, a **Cisco level** system might be the way to go. From my experience, 3CX also seems to be a good fit where there is on-site IT to watch over and take care of telephony issues.

3CX is not a turnkey hardware phone system

What gets most small businesses interested in 3CX is the low cost and do-it-yourself possibilities. Also the open architecture lets you use nearly any standard SIP hardware and this is really appealing. After working through several interoperability issues, we might start wishing for a turnkey proprietary phone system "that just works," and 3CX is a system that needs to be integrated.

3CX is not done

In fact, 3CX is being developed at a dizzying pace. In just a few versions, the web server was changed from Apache to IIS, the user interface completely redone, a new softphone was added, the Call Reporter was rewritten from Microsoft Access to a self-contained Windows application, a Hotel add-on was added, the 3CX Assistant was added, and lots of new features keep pouring in. 3CX is certainly a work in progress and a moving target. It's not at all uncommon for updates to come out several times a month. To keep an eye on these developments, you can follow the URL: http://wiki.3cx.com/change-log.

3CX does not have "key system" replacement features

The easiest way to explain a **key system** is to give an example. A call comes in on line 1, and John picks up the phone. The caller wants to talk to Joe, so the call is put on hold. John tells Joe to pick up the phone and Joe presses line 1 and says "Hello." 3CX uses the call park and call transfer paradigm instead, which works well but is sometimes a hard feature to give up for users who are used to a key system. The IP PBX feature that allows putting the call on hold and the other person picking it up by pressing the "line" button is sometimes called **shared line appearance**.

3CX integration with Microsoft Office Communications Server is not supported

Microsoft Office Communications Server (OCS) is a communication server that started as an instant messaging server. It has grown to have voice and collaboration features and is often integrated with an existing phone system. As Microsoft appears to be grooming OCS to become a full-blown communication system capable of replacing a phone system, 3CX has made the choice to not support integration with OCS. 3CX does integrate with other instant message-only servers (such as Openfire) easily and nicely. While this is not officially supported by 3CX, you will find more help at 3CX support and forums.

3CX currently does not have the ability to do multi-tenant

Multi-tenant is the ability for multiple companies to use one install of 3CX. 3CX's current answer to multi-tenant needs is **virtualization**.

3CX does not do multiple languages simultaneously

This may be needed for countries that need to support prompts in more than one language at a time. 3CX can do many languages, but only one at a time. So, if you need a digital receptionist prompt to say *Hello, for English press 1 and French press 2*, and then all prompts after that will switch to the language you selected, remember that 3CX does not do that.

Summary

We should now have a full understanding of the 3CX Phone System. In this chapter, we compared hardware and software phone systems, as well as 3CX and Asterisk phone systems.

We quickly took a tour of the major components that make up 3CX. The phone system itself (made up of 13 services), the 3CX VoIP Client, the 3CX softphone, the 3CX Assistant, the Call Reporter, and Hotel module.

We reviewed the characteristics that really define 3CX. Its ease of use, its ability to interoperate with many standard SIP hardware vendors, and that it is unabashedly a Microsoft Windows product are just some of those characteristics. We also took a look at the fact that 3CX is not on the same level of maturity as a Cisco phone system. It's not a turnkey phone system that you can just plug and play. Finally, it's not a finished product; development continues at a dizzying speed.

We noted some features that we shouldn't expect 3CX to provide, such as key system features, Microsoft OCS integration, multi-tenancy, and multiple languages at the same time.

In the next chapter, we will be taking a look at the items we should get together to set up a 3CX Phone System and what computer or server we should use for our system. Then we'll download 3CX and finally go through the 3CX install and a basic test of the system.

2
Downloading and Installing 3CX

Now we can get down to the business of actually getting 3CX installed. We'll take a look at what we will require in order to set up a 3CX Phone System, what computer or server we will need to use for our system, where and how to get 3CX and the keys to use it, and finally the 3CX installation. In this chapter, we will have a look at the following:

- What you will need
- 3CX server requirements
- Choosing a Windows operating system
- Downloading 3CX and getting a key
- Installing the 3CX Phone System
- Checking if 3CX is up and running

What you will need

There are just a few things that you will need in order to build a complete phone system using 3CX, which are:

- A Windows desktop or server computer
- Some DHCP server (a small router will do)
- Three hardware handsets or softphones
- Broadband Internet and a SIP trunk
- A SIP PSTN Gateway device (if we want to connect to PSTN lines)

Why three phones? It is because three phones will let us test all the different call transfer and call park scenarios ourselves. For example, phone #1 can call to phone #2. Then phone #2 can transfer the call to phone #3. With three phones you can basically test any combination that can occur in the real world. If you can't afford hardware phones, software phones will suffice.

As it is possible to run 3CX in a virtualized operating system, it is possible to set up a complete phone system without buying any hardware—run the 3CX Phone System on a virtual server, install softphones on laptops you already own, and use a SIP trunk to connect to the outside world! Wow! My suggestion though is that for your first install, get a real server or desktop PC. This will be one less thing to worry about.

We'll talk about the requirements for our system in just a moment, but let's quickly talk about your phone handsets and PSTN gateway. Remember that, while 3CX is open and can integrate with any SIP compliant hardware, using 3CX-supported hardware will make your life a bit easier and that is what I recommend. The official list of 3CX-supported hardware can be found at: http://www.3cx.com/sip-phones/index.html

You will need to select your phone handset very carefully because a good or bad handset can make or break your phone system implementation. Often, because handsets are a large part of the investment of a phone system, handsets are selected on the basis of price only. However, don't do this, no matter how good 3CX is, if your users work with a phone handset that is quiet, doesn't work right, cuts out, has buttons that don't work correctly, or has a low speakerphone volume, you'll get bad reviews.

My suggestion is to get 3CX up and running successfully before trying to get unsupported hardware working. It really is a separate task.

For selecting a PSTN gateway, I recommend getting a 3CX-supported unit. The gateway device is probably the most complicated device to get working just right on your phone system, so don't skimp here. 3CX supports the most well known and often used gateways. Once again, here is the URL: http://wiki.3cx.com/gateway-configuration

Some people ask "Which is your favorite phone handset and PSTN gateway?" Well, remember everyone has their own preferences. I lean heavily toward products that work without much fiddling and are very stable. Price comes somewhere later in my priority. What's my favorite? I go for Snom phones and Patton gateways.

Your 3CX server hardware requirements

What are the hardware requirements for 3CX? It might be astonishing but for a long time 3CX (the company) did not release specific requirements. The recommendation was "a currently sold Windows desktop will do." This might seem to be a somewhat "hazy" recommendation, but I've found that this rule of thumb actually can work quite well for a 10 to 15 extension system.

The first time I installed 3CX, I installed it on a fanless 1GHz VIA appliance PC with 512MB of RAM and a 100GB hard disk drive. I thought that, because it was a phone system, it wouldn't take many resources. Well I was wrong. When calls would come in, there was a long delay before the digital receptionist would pick up in other places when prompts played to the caller. I fiddled for quite a while before I moved the phone system to a modern desktop computer and then all the problems went away. With Asterisk, people often say you can turn an "old unused pc" into a phone system. 3CX is a Windows-based system, so make sure that you have a sufficiently powerful machine.

3CX now publishes minimum hardware requirements that appear during installation, which are as follows:

- 1GB RAM
- Pentium IV processor or higher
- Audio playback capability

> 3CX uses an **MP3** file residing on the 3CX server as the Music on Hold source. At this time, plugging an audio source into the audio card's IN port is not a feature. Also interfacing to PA systems by plugging into the audio card's OUT is not supported. We will talk more about PA system integration in Chapter 8.

Obviously, if you are running 3CX alongside an **Exchange Server, Active Directory**, and **File Sharing** on a **Microsoft Windows Small Business Server**, you will need a more powerful machine. 3CX has published a test in such a situation on their official 3CX blog at `http://www.3cx.com/blog/voip-howto/no-dedicated-server-needed/`.

For the test, the server was configured as follows:

- Intel Core 2 Duo E4500 2.20GHz
- 4GB RAM
- 50GB SATA hard disk drive
- 100MBps network connection

In the test, **Windows Small Business Server** was running **IIS, Exchange Server,** and **Active Directory**. 3CX was installed using the **Cassini** web server option. A load equal to 25 users heavily using Exchange Server was put on the server. Now a 16 simultaneous calls load was put on 3CX. In one hour, 2,000 calls were made. During this test for all 3CX services, the CPU usage was at less than 15%. The 3CX memory usage was approximately 300MB with Cassini using about 100MB. Exchange Server did not use more than 10% of CPU with a total CPU usage of about 30%.

Choosing a Windows operating system

3CX will run on most business editions of Windows, but there are some considerations for the different editions to take note of:

- **Windows XP Professional**

  I personally think that Windows XP Professional makes a nice small office 3CX Phone System server and is a great choice for a first 3CX install. It can be clean, lean, and mean. It is especially compelling when you have a network and have another server taking care of DHCP.

> For performance reasons, Windows XP Professional along with the Cassini web server should be used for systems with fewer than 20 users. Consider upgrading to a server operating system when you reach this number of extensions.

- **Windows Vista Business, Enterprise, and Ultimate**

  If you are using Windows Vista, make sure that your machine has plenty of resources and has **User Account Control (UAC)** turned off. Also, see that the firewall doesn't get in your way.

- **Windows Server Standard 2003 and 2008**

  This may be the best fit for a 3CX server in a slightly larger installation. Like SBS, it will include a robust DHCP server for provisioning phones using **DHCP option 66**.

> This is just in: Microsoft has just released another version of their Windows Server line called **Windows Foundation Server**. This server will have all the features of Windows Server Standard, be priced at Windows desktop OS price levels, and be limited to 15 SMB connections. In my opinion, IP PBX administrators will want to keep their eye on it!

In fair disclosure I need to admit that I personally favor having discrete servers or servers doing one job. In an SBS environment it's so easy for different server applications running on the same server to conflict, whether it is over RAM or ports, or whatever. With a real-time server like a communication server, even a little delay, stutter, or chop can cause a user to give the phone system administrator a support call. However, remember this is a personal opinion.

3CX is also supported on Windows Server 2008. Still, you'll need to ensure that the Desktop Experience role is enabled. Without this you will have no audio prompts, and this will make your phone system useless.

- **Windows Small Business Server 2003**

   Windows Small Business Server has a lot going on already, such as file sharing, Exchange Server, and Active Directory, so it can be potentially tricky to get yet another server running on it. However, it can be done, and 3CX will support it, too. My suggestion for a first install is to try something simpler like Windows XP or Vista.

Starting with a clean operating system install

For our first 3CX server, we will need to start with a freshly installed and clean operating system. 3CX is quite a complex application as there are 13 services running, and it can use several TCP ports and audio interfaces with the PC's hardware audio capabilities. With 3CX, there is quite a lot going on, so a clean install will make our lives smoother.

Some real life experience: The first time I installed 3CX, it was on an older Windows XP Professional computer that had a lot of test applications installed and uninstalled on it. I had so many problems that I was about to write off 3CX as not ready for primetime when, on a hunch, I decided to install 3CX on a clean Windows XP Professional computer. The install was flawless and everything just worked! After working with other 3CX installations, I've found that 3CX likes a clean operating system and is perhaps a little more sensitive than some other applications.

Getting the Microsoft stack in place

The 3CX Phone System is largely based on several Microsoft Windows technologies that are stacked on top of each other.

You don't need to understand all the details about each layer; all you need to know is that each one needs to be installed and working properly. The Microsoft "stack" used in 3CX is as follows:

- Windows operating system (and updates)
- .NET Framework 2.0
- IIS (Internet Information Services, unless you use Cassini)
- ASP.NET

Note that I said 3CX is "largely" based on Microsoft technologies. The 3CX database uses the PostgreSQL database engine and not Microsoft SQL. There is no problem with PostgreSQL, but many Windows users may not find it as familiar as Microsoft SQL when attempting to integrate with it. For example, you cannot connect to it from Microsoft Access as simply as Microsoft SQL.

Downloading 3CX and getting a key

To get to the 3CX Phone System download page, visit:

`http://www.3cx.com/phone-system/download-phone-system.html`

Fill in your information with a valid e-mail address, and you will get a demo key that will allow you to try out all the commercial features of the 3CX Phone System. You will also get a link to a nice PDF manual for 3CX, or if you are very security conscious, you can click on the link under the heading **Updates**, which will take you to the links to download without giving you a demo key. You can also go directly to the following link:

`http://www.3cx.com/phone-system/downloadlinks.html`

3CX is approximately a 40MB download. You may also want to download the **3CX Assistant** and **VoIP Phone** right away from the same page.

Free key versus a two-user test key

If you filled in a valid e-mail while registering, to download 3CX you will receive a demo key by e-mail that will allow all features of the 3CX Phone System Commercial edition. The limitation of this key is that it will allow *only two simultaneous calls*.

To activate the two-user demo key, you will need to click on **Settings | Activate License**. To make things simple, I suggest you copy and paste the key from the e-mail that you received.

If you decide that you want to go back to the 4 simultaneous calls Free edition, your can simply click on the **Free Version** button and the 2-call demo key will be removed.

Once again, most people find that to thoroughly test phone features, you really need to have three phone extensions. At this point, 3CX only provides a 2-call demo key. I can understand 3CX's position that a 3-call key could be used for real use. A 3CX reseller can be a good resource to help you test features using a demo system they may make available to you.

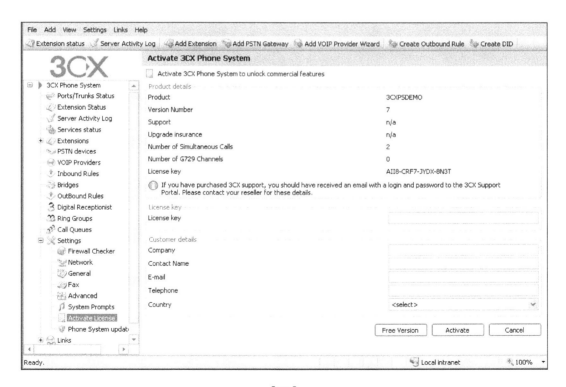

For a neater comparison between the Free and Commercial editions of 3CX, visit the following URL: `http://www.3cx.com/phone-system/edition-comparison.html`.

Starting the install

Now we will get started on installing 3CX, which we will do by clicking on the install file that you previously downloaded. Usually, this file is named something like `3CXPhoneSystemXX.exe`, where `XX` is the version number.

The requirements screen

The very first screen you'll see when installing 3CX is the requirements screen, as shown in the following screenshot. We've already talked about these requirements, but let me stop here and emphasize that these are *requirements*. If you proceed and one of these requirements is not met, you will run into problems moving ahead. Click **Next**:

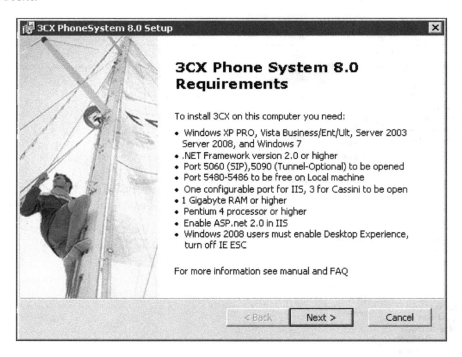

The recommendations screen

The next screen that we will see is the recommendations screen. If you haven't met all of these recommendations, you will still be able to continue. To make life easier, I would use Internet Explorer. Using **Cassini** instead of **IIS** might actually be simpler for a small install. Click **Next**:

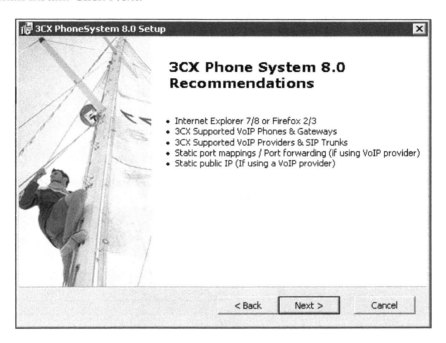

The EUL Agreement

The **End-User License Agreement (EULA)** screen reminds us that 3CX is not **Open Source**. The EULA looks quite normal to me, but you will need to read it and make sure you agree with it before clicking **Next**:

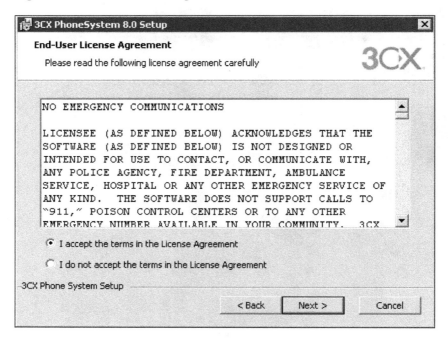

 One part of the EULA that you will want to take note of is that 3CX does not support 911 or emergency call handling. Make sure you check whether your organization requires 911 handling or not.

The install folder screen

In order to install 3CX efficiently, you will need to select where you want it installed before clicking **Next**:

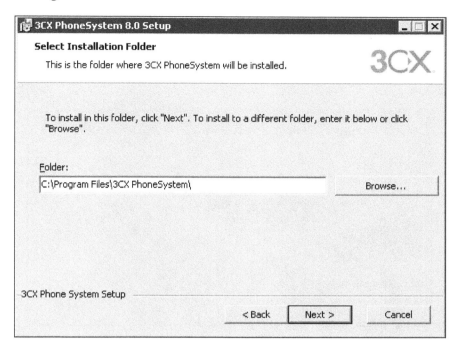

Selecting IIS or Cassini web server

You will also need to select a 3CX web server — you can select Microsoft IIS or Microsoft Cassini. Cassini's performance will not be as good as IIS in larger installations, but Cassini is fine for a test system or small office phone system with 10 to 15 extensions.

 If you are unsure which web server to select, I suggest you use **Cassini**. In Microsoft Small Business Server environments, this will avoid configuration issues that can occur with **Outlook Web Access** (OWA), **SharePoint**, and other applications that use IIS.

After you have selected which web server you want to use, you can click **Next** as shown in the following screenshot; and 3CX will be installed. This will only take a minute or two. When it is done, we can click the **Complete** button that will launch the **3CX User Settings Wizard**:

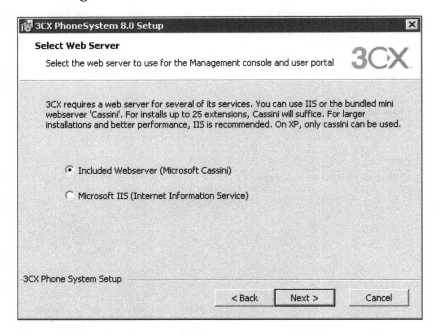

If you want to use IIS, it must be installed and properly configured before you try to make 3CX use it.

The 3CX User Settings Wizard

The **3CX User Settings Wizard** will automatically guide you through getting your 3CX Phone System set up in a basic way. It will start with some general or global 3CX settings and then move on to add some phone settings, such as extensions, which is the operator extension, and then set up a gateway. Here are a few more settings:

- **Language**: This will select the language for the prompts played in the system.
- **Settings**: Whether you are creating a new 3CX or restoring a backup, if you are reinstalling 3CX, you can choose to restore a backup instead of going through this wizard again.

- **Extension Digits**: Lets you select "3," which is a very common setting because it allows a reasonable amount of extension numbers.
- **SIP Domain**: We can use the IP address of the 3CX server.
- **Mailer Server**: We select the SMTP server that will send e-mail for this 3CX server.
- **Administrator Login**: We can choose the administrator's username and password. It is case sensitive, so we need to make sure that we remember what it is.

 All of the 3CX general settings can be changed later too, except SIP domain and extension digits, so you don't need to agonize over these settings.

The following screenshot shows the first screen in the **3CX User Settings Wizard**:

Creating user extensions

Next, we will need to set up some phone extensions using basic extension settings. All we need to enter at this point is an extension number, first name, and last name. If we enter an e-mail address, an e-mail will be sent to notify the user of his/her extension being created. The e-mail address will also allow voicemail recordings to be sent to the extension user. Let's set up 3 extensions using extension numbers **101** to **103**.

When you add an extension, the extension number will be used as that extension's **PIN** and **password**. In a live system, you will likely want to change this to tighten up security.

Following is the **Add User Extension** screen from the **3CX User Settings Wizard**:

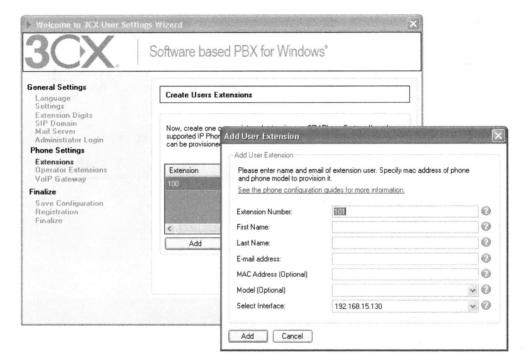

Operator extension

After we have added several extensions, we will need to click **Next** and set the operator extension. The operator extension is a special extension for two reasons:

- It is the default extension to which incoming calls will be directed
- It is the extension to which calls are routed, when a caller wants to "breakout" of a voicemail prompt

Click **Next** through **VoIP Gateway**, and now the database that 3CX uses to store its setup and configuration details will be created using the settings that we just entered. Depending on the speed of your server, this will take upto several minutes. The wizard will inform you of the progress as it works on creating the database. Once the database is created, clicking **Next** will take us to the **Registration** screen:

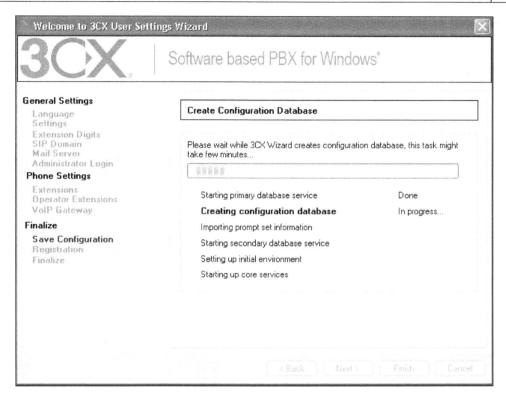

Registration

The **registration screen** will allow us to type in our contact information, but this isn't required for the free version, so we'll just click on **Skip**. We are now done with the **3CX User Settings Wizard**, we can now click **Finish,** which should launch the **3CX Management Console**.

Logging in to 3CX for the first time

Now the 3CX Phone System is completely installed, and we can log in to see if we have everything just right. We can do this by browsing to `http://localhost:5481` directly from the computer we installed it on (if it did not automatically launch when we clicked **Finish** on the **3CX User Settings Wizard)**.

If we had used an IIS web server instead of Cassini, the URL would have been slightly different.

I suggest we try logging on to our 3CX server directly from the console first. We will also be able to log on from a remote computer by substituting `localhost` with our 3CX server IP address or computer name. Here is an example: `http://192.168.1.10:5481, http://3cxhome:5481`.

 When logging on to the 3CX administrator console, we will need to make sure that we provide the username and password in a case-sensitive format.

If everything is working fine so far, we should see the 3CX administrator console login screen. We will provide the username and password that we selected earlier and click **Login**.

 When we installed 3CX, we selected what language the audio prompts will be in. When you log into 3CX, you can select what language the web pages are presented in by selecting your language of choice from the drop-down list directly above the login username and password, as shown in the following screenshot.

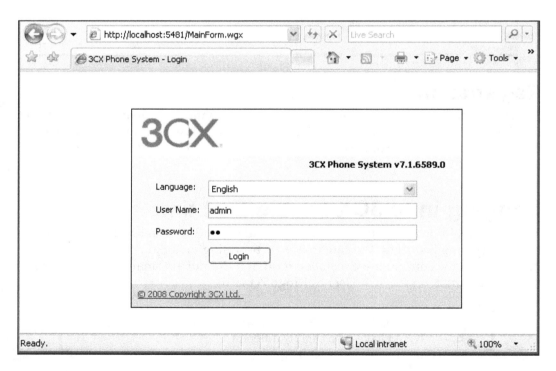

You should now be logged in to 3CX.

Checking the status of 3CX

We can now click on **Services status,** as shown in the following screenshot, to check that all system services of 3CX have started and are running correctly. If this is the case, then we have successfully installed the 3CX Phone System:

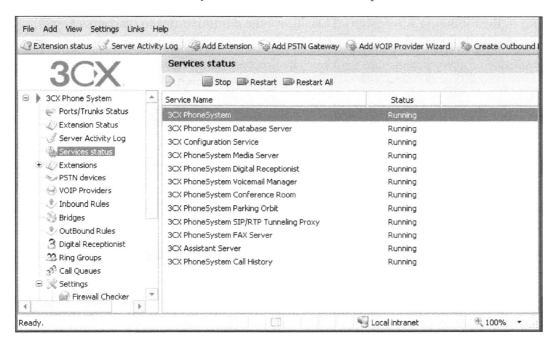

We can also take a look and see if our extensions were added correctly by clicking on **Extension Status** shown in the following screenshot. The **Extension Status** screen is an excellent place to view the status of all your extensions. We can see that our extensions were added, but we can also see in one quick glance the status of each extension:

- **Red**: **Not Registered** means no phone is registered to this extension. (This is how your system should look as we didn't connect any phones yet.)

- **Green**: **Registered** means a phone has been connected and registered to this extension.

- **Yellow**: The phone connected to this extension is on a call.

If an extension is on a call, the **IN/OUT** column will show whether this is an incoming or outgoing call, the **Caller ID** column will show the caller ID of the remote caller, and the **Destination** column will show what trunk was used to connect to the remote party.

Also, note the **Disconnect Call** button above the **Status** column (disabled/grayed out in the following screenshot). This allows the administrator to disconnect an ongoing call if needed:

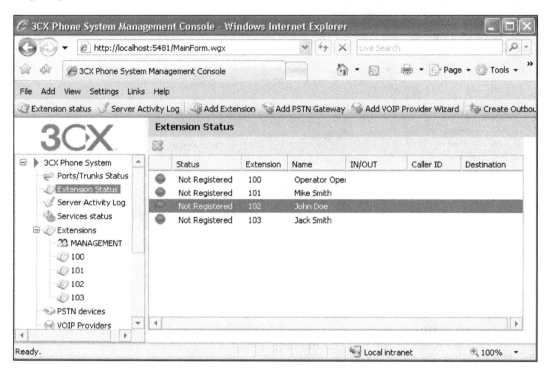

Now, let's click on the **Server Activity Log**. It allows us to see detailed logs about live events on the 3CX server. The **Refresh** button adds new events since the last refresh. You can click **Copy Selected Text** to copy some of the lines to another program. The level of logging reporting on this screen can be changed by going to **Settings | Advanced | Logging level**. The logging level can be set to **Low**, **Medium**, or **Verbose**. You might be tempted to set the logging level to **Verbose**, but remember this will dramatically increase the load on the server and is only meant for temporary debugging situations. The following screenshot shows the **Server Activity Log**:

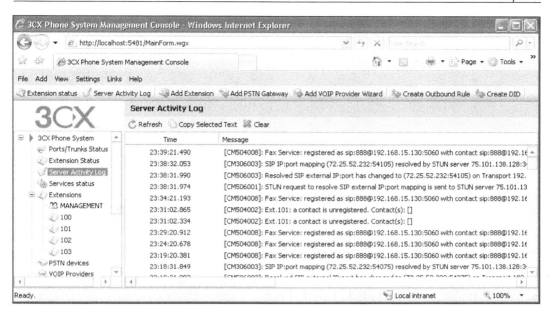

 As we have been navigating the 3CX administrator console interface, you may have noticed a small lag in response when clicking on options. Please note that this is normal. When 3CX moved to the ASP.NET user interface, I noticed there was some normal delay. Sometimes the delay can be long enough that you might think nothing is happening, but just check the progress indicator on the right side of the screen.

Summary

We should now have a full understanding of what is required in order to build a 3CX Phone System. In this chapter, we looked at the hardware and operating system requirements and how to get the 3CX download at `http://www.3cx.com/phone-system/download-phone-system.html`. Finally, we learned how to install the 3CX system, making sure that things were running correctly by checking up on 3CX's status.

In the next chapter, we will connect softphone and hardphone extensions to 3CX and test to make sure they are working. We will log in to the extension portal to see the status of the extension and learn how users can change some information about themselves. We will also explore some of the more advanced extension configuration options, such as voicemail and call forwarding rules.

3
Working with Extensions

A phone system without any extensions doesn't make much sense. In this chapter, we'll take a look at devices that can connect to a 3CX Phone System as extensions. Next, we will make sure that we can connect to our 3CX server from another computer, and then we will configure, connect, and test some software and hardware extensions. After this, we will look at the MyPhone UserPortal, which allows users to view and change their own extension settings. We'll move on to advanced extension settings, extension groups, and wrap up by looking at multiple extension editing available with the 3CX Phone System. We will look at:

- Devices that can connect to 3CX as extensions
- Ensuring connectivity to our 3CX Phone System from another computer
- Configuring some software and hardware phones
- The MyPhone UserPortal
- Extension settings and groups
- Editing multiple extensions at the same time

By the time we are done with this chapter, we will have a phone system that is configured to do internal calling and all that will be left is to connect our private phone system to the outside world.

Devices that can connect to 3CX as extensions

Earlier we noted that traditional phone systems depended on the manufacturer of the PBX phone system to make the extensions that connected to it. With the introduction of the SIP protocol standards, any VoIP phone system that adheres to SIP standards can connect to any SIP-based extension that adheres to SIP standards. This is a powerful openness that the 3CX Phone System leverages to allow many phone handsets and softphones to be used as extensions with 3CX.

 This wonderful ability for all SIP devices to interoperate is theoretical. Sometimes the SIP standard has multiple ways of doing something and different manufacturers use different methods making things break down. Thankfully, 3CX has done a lot of interoperability testing which is available at `http://wiki.3cx.com/phone-configuration/firmwares-tested`.

Let's take a look at some different categories of extensions that can connect to 3CX.

Softphones

We've already covered the 3CX Phone. One nice thing about softphones is that they have a very low cost as compared to hardphones and they often bring a higher level of integration to your computer. There are some other well-known manufacturers who also make some nice softphones.

X-Lite by CounterPath

X-Lite is a very nice, free softphone that adds some features not available in the 3CX softphone. One such feature is **video phone** conversations between extensions on your phone system. 3CX does not officially support videos, but since X-Lite adheres to protocol standards, the 3CX Phone System will pass the video along. Another feature of X-Lite is a built-in instant messaging client that works with a standard SIP simple instant messaging server. This phone is supported by 3CX and more information about it is available at `http://www.counterpath.com/`.

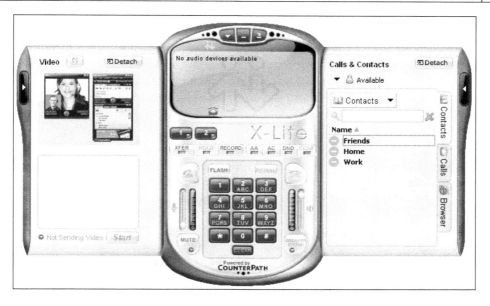

Zoiper Communicator

The Zoiper Communicator is another SIP softphone that has a free edition, a nice user interface, and brings some interesting features within the reach of 3CX. It can act as a **T.38 Fax** sending and receiving device, has video calls ability, has a built-in instant messaging client, and provides a hosted instant message server too. Zoiper is *not* on the supported list of softphones for 3CX, so you will be on your own if you use it. More information about the Zoiper Communicator can be found at `http://www.zoiper.com/`.

SIP phones

Hardphones or handsets are similar to what has been on desks for the last 50 years and are still the most popular way to use a phone system. Some very common SIP phone manufacturers are Polycom, GrandStream, Snom, Cisco, Aastra, and Linksys. Various phones from these vendors are on the 3CX supported list.

Analog phones

VoIP phone systems are amazing at accommodating old hardware. Using an **Analog Telephone Adapter (ATA)** device that converts an analog phone signal to the SIP protocol, you can use almost any old analog phone device with your modern, state-of-the-art VoIP phone system.

Other SIP hardware and software devices

Following are some other SIP-based hardware devices that can act as an extension to 3CX:

- **Ceiling speakers that can be used to page**:
  http://www.cyberdata.net/products/voip/digitalanalog/
  ceilingspkr/index.html
- **Security camera you can call to listen to, talk to, or watch (via video phone)**:
  http://www.mobotix.com/eng_US/content/view/full/1551
- **Intercom door phones**:
  http://www.cyberdata.net/products/voip/digitalanalog/
  intercomindoor/index.html
- **Door opener**:
  http://www.abptech.com/products/its.html
- **Software to do outbound calling**:
  http://www.nch.com.au/in/voicemail.html

This is not an attempt to list all the SIP devices that could possibly connect to 3CX, but just to give you an idea about what some of the possibilities are. Also note that these devices are not on the recommended 3CX devices list.

Verifying basic network connectivity to our 3CX server from another computer

Now we want to move on to connecting a phone extension to our 3CX Phone System. For our first extension, we'll use the 3CX VoIP Phone.

I suggest that we first test if we have connectivity to our phone system from another computer and ensure that no basic network problems exist. Ping your 3CX server from the PC on which you want to install the 3CX VoIP Phone. If that succeeds, we can log on to the 3CX administrator console from the remote computer to make sure that we can connect at that level.

 If you cannot ping the 3CX server then you have a basic network problem. A very common issue is that a firewall is running on the 3CX server or the softphone PC. Make sure all firewalls are turned off to continue.

Now let's make sure we can log in to 3CX from a remote computer by opening a browser and going to `http://<ip_address>:5481`.

You will want to replace the `<ip_address>` with your 3CX server IP address. If you can successfully log in, we should be ready to set up an extension and connect it.

 At the moment, administration of the 3CX Phone System is only supported from Internet Explorer and Firefox.

Basic extension setup in the administrator console

We added extensions during the **User Settings Wizard**, but now we will add an extension using the system management console. There are three ways to add an extension as shown in the following screenshot:

- **Navigation pane**: Click **Extensions | Add Extension**
- **Drop-down menu**: Click **Add | Extension**
- **Quick launch toolbar**: Click **Add Extension**

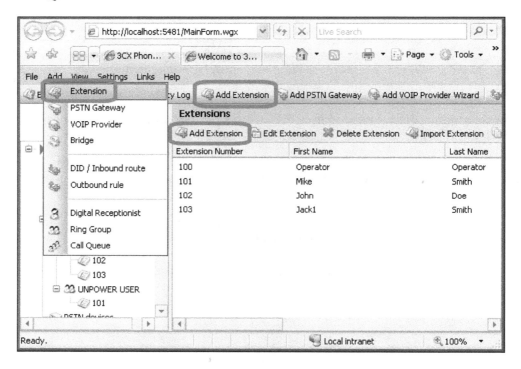

We will now be presented with the extension's **General** settings tab. You will notice that the **Extension Number** is automatically populated with the next available extension number. Because we entered 101 to 103 in the tutorial, we are presented with **104** in the **3CX User Settings Wizard**.

First and last name

First Name and **Last Name** are fairly straightforward but remember that the **Last Name** is used when incoming callers use the **Dial by Name** feature in 3CX.

ID, password, and pin

Let's also notice that the **ID**, **Password**, and **Pin** have been assigned **104** automatically, for us to save time. Unlike in the settings wizard, we have the ability to change those passwords immediately to be more secure. For the sake of time, let's just leave them as they are.

E-mail address

Enter a valid e-mail address for this extension user. If the **Notify User When Extension is Added** setting is turned on, an e-mail can be sent to the user informing them that we have set up an extension for them. (This is a nice way to welcome a user to the phone system!) The e-mail address is also used if you set voicemail to be delivered to e-mail.

 The default message sent to a user when an extension is set up can be modified under **Settings | General | Mail Server | Customize Mail Notification**. This is the same place you would turn on **Notify User When Extension is Added**, by default, it is turned off.

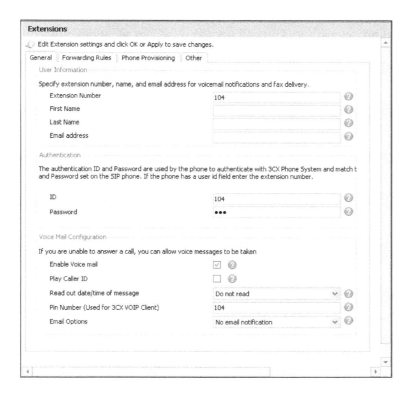

Voicemail configuration

If you don't want this extension to have voicemail, just uncheck **Enable Voice mail**. The **Play Caller ID** and **Read out date/time of message** are settings related to how the 3CX voicemail is read back to you. Let's just keep these settings turned off.

You can also select what happens when you get a voicemail. Should the system not bother you at all? Should it send an e-mail saying you should check voicemail? Should it actually send the voicemail along with the e-mail as an attachment? Or, should it go the whole way by sending an e-mail attachment and then even deleting it in 3CX? It's up to you to decide.

Now we are done with the **General** setup of the extension.

 You need to click **Apply** before clicking on the **Forwarding Rules** tab or the extension you just created will not be available in drop-downs to make your forwarding rules.

Forwarding rules

We could click **Apply** and be done at this point, and we would have an extension setup so that one extension could call another one, but we'd like voicemail to kick-in, so let's click on the **Forwarding Rules** tab. We'll spend more time on **Forwarding Rules** later but, for now let's just set up one simple rule as shown in the following screenshot. Click **Apply** when you are done:

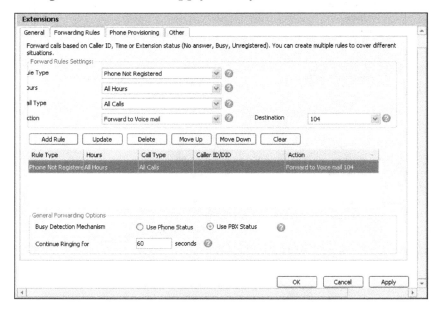

The simple forwarding rule that we just set up will send a caller calling this extension to voicemail if no phone is registered to it.

 Settings do not take effect until you press **Apply**. You will want to get into the habit of clicking **Apply**.

We are now done with a simple extension setup and we can get a softphone installed and registered.

Installing and connecting the 3CX VoIP Phone

We'll use the **3CX VoIP Phone** softphone as the first phone that we will connect to our 3CX Phone System and we'll use a PC or laptop as our first extension. You can download the softphone from the following URL and you can just click on it to install:

```
http://www.3cx.com/phone-system/downloadlinks.html
```

Simply click **Next** on each screen, and when the install is done the **3CX VoIP Phone** will be launched. The **Connection settings** screen will open, as shown in the following screenshot, to configure the 3CX VoIP Phone for the first time.

We will set up this extension to be 101. There are only four fields you will need to fill in:

- **Extension: 101**
- **ID: 101**
- **Password: 101**
- **I am in the office – local IP**: The IP address of your 3CX server

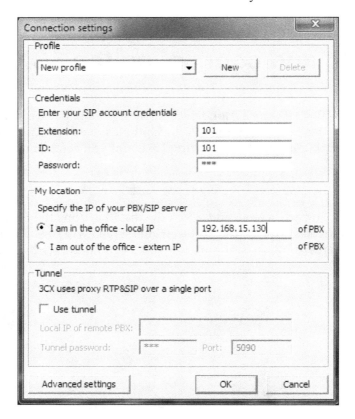

When you click **OK**, the softphone will connect to 3CX. You may get a Windows firewall message as shown in the following screenshot, make sure you allow 3CX VoIP Phone access to private networks. The status area of the 3CX VoIP Phone will say **Connecting...** momentarily, then it should stay at **On Hook**.

Testing the extension we just connected

We now have our softphone registered and ready to use. There are several ways to test if an extension has registered correctly but usually I am not real academic at this point. A very simple, pragmatic test is to make a call. As we don't have any other extensions registered, we can just call 999 (or 99 if we have a two-digit extension setup or 9999 if this is a four-digit install), which is the default **dial code** to connect to the voicemail for this extension. If we hear the prompt *Please press personal identification number and then press pound*, then we have successfully connected the 3CX VoIP Phone to our system.

Checking that system console indicates the extension as registered

Another way to visually see if a set of extensions has correctly registered is by the **Extension Status** screen in the system management console. A red light indicates no phone is registered, a green light indicates a phone is registered, and a yellow light indicates a registered phone is on a call.

Testing that we can call another extension

Another way of testing if an extension is working correctly is to see if we can call from one extension to another. You might be thinking "Wait a minute, there is only one phone connected to this system!" As we made a forwarding rule for extension 104 that sends callers to extension 104's voicemail, we can call that extension even if there is no phone connected! Let's go ahead and call 104. We should get a prompt *Record your message and press pound*, verifying that calling another extension also works. The following screenshot shows the 3CX VoIP Phone as it should appear when correctly registered to 3CX Phone System:

Connecting a Snom 360 phone

Connecting a Snom 360 phone to the 3CX Phone System is fairly straightforward. You log into a web interface and configure it much like many popular routers. Unlike the 3CX softphone, you have the ability to auto provision the Snom 360 phone as well as quite a few other popular phone handsets. **Auto provisioning** allows you to push a configuration to the phone instead of logging into each phone that you connect to your phone system. Auto provisioning will speed up things if you have several phone handsets.

To set up auto provisioning, we will go back to the 3CX **Edit Extension** settings page for extension 102 and click on the **Phone Provisioning** tab, as shown in the next screenshot. The two most important settings we need to provide for provisioning are the **MAC Address** and **Model** of the phone. The **MAC Address** of the Snom phone can be acquired by looking at the white label on the bottom of the phone or by pressing the **Help** button on the phone. The phone model can be selected from a drop-down list.

We can also configure what extensions the **Busy Lamp Field** (**BLF**) lamps on the Snom phone monitors by selecting the extension from a drop-down list. After you are finished with the BLF configuration, press **Apply**. Now we are done in 3CX and will move to what we need to do on the Snom phone.

> A common question: Can the BLF lamps show the status of a phone trunk in 3CX? The answer: No, 3CX does not currently provide the ability to monitor phone trunks using BLF lamps on the phone handset. You can see the trunk status via the 3CX Assistant.

> The free version of 3CX does not enable BLF on phones. So even if you set up BLF in the provisioning tab, the BLF lights will not light up when that extension is on the phone.

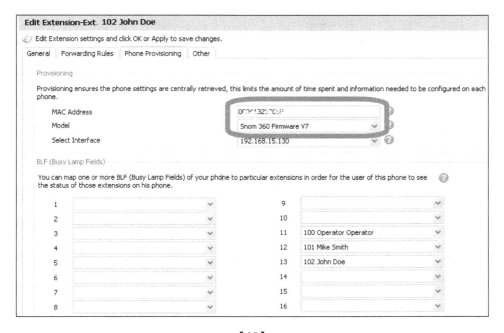

As shown in the previous screenshot, the phone provision tab in 3CX allows you to configure which phone extensions each BLF lamp on the Snom phone monitors.

Power up the phone and it will, by default, get an IP address from the DHCP server on your network. The Snom phone will display the IP address it received while booting, but if you miss it when it booted you, press the **Help** button on the phone and the IP address and MAC address will be shown on the display.

> The Snom 360 is a very common SIP phone that can be used with 3CX. Some features that set it apart are the built-in 12 BLF and buttons, a record button, Power over Ethernet, and business quality construction.

To begin configuring the Snom 360 we will log in to its web interface by typing the Snom IP address into the Internet browser: http://192.168.1.22.

The default password for Snom phones is **0000**. You should be greeted by the Snom **Welcome to Your Phone!** page after you log in, as shown in the following screenshot. The Snom welcome page allows you to dial phone numbers, and see the dialed, missed, and received calls right from a single web page. On the left you can navigate to other screens too.

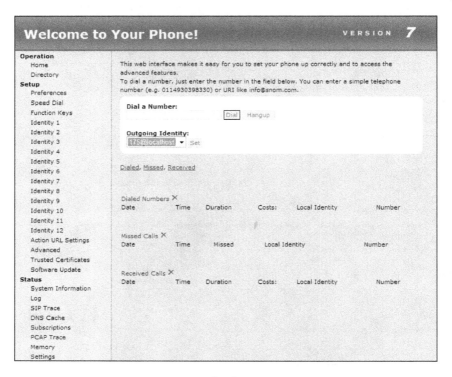

The first thing we will want to do if this is not a new phone is reset the phone to factory defaults by going to **Advanced** and then click on **Reset**. This will reboot the phone and erase all settings.

Next we will want to make sure we are on an appropriate firmware version for this phone model as per the following 3CX web page:

```
http://www.3cx.com/sip-phones/index.html
```

We will click on **System Information** to verify that this Snom phone has the 3CX recommended firmware, which is **7.3.14** at the time of writing this book. If your phone does not have the appropriate firmware, you will want to update the firmware. As the phone in the following screenshot has firmware version 7.1.35, it needs to be updated.

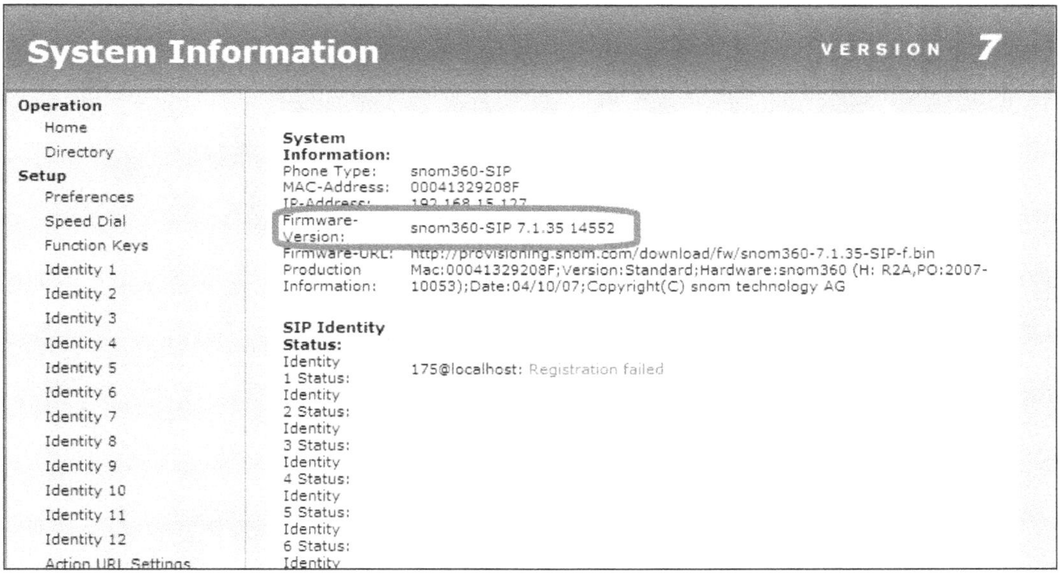

The **System Information** page shows important information about the Snom phone, such as **MAC-Address**, **IP-Address**, **Firmware-Version**, and whether the phone is registered. This is a great place to make sure your firmware is up to date.

If your firmware needs to be updated, click on **Software Update**. Fill in the URL `http://provisioning.snom.com/download/fw/snom360-7.3.14-SIP-f.bin` in the **Firmware** field and click **Load**. Your Snom phone will now reboot. You may need to press **OK** on the phone to start the update.

As firmware is changing all the time you can find the latest Snom firmware at `http://wiki.snom.com/Snom360/Firmware`. You can be pretty safe by just substituting the firmware numbers in the above URL with the ones recommended by 3CX. An example firmware location on Snom's site is `http://provisioning.snom/download/fw/snom360-7.3.30-SIP-f.bin`.

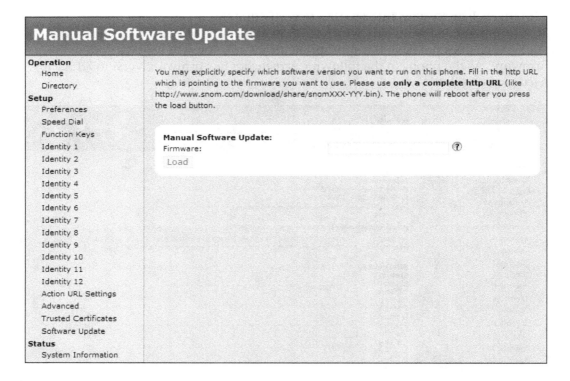

After you have updated the firmware to the 3CX approved version we are ready to type in the provisioning information. Click on **Advanced / Update** tab. On this screen, the two key pieces of information are the **Update Policy** and the **Setting URL**. The **Update Policy** should be set to **Update automatically**. The **Setting URL** should be pointed to the 3CX Phone System provisioning file for this extension. The **Setting URL** should look like: `http://192.168.15.130:5481/provisioning/cfg{mac}`

Only replace `192.168.15.130` with the IP address of your own 3CX Phone System. You can press **Save** and then **Reboot** and then **Yes**. When the Snom phone reboots it should be provisioned and registered with your 3CX system. The **Update** tab of the **Advanced Settings** screen is where the **Setting URL** is configured. This URL points to the provisioning file for this phone:

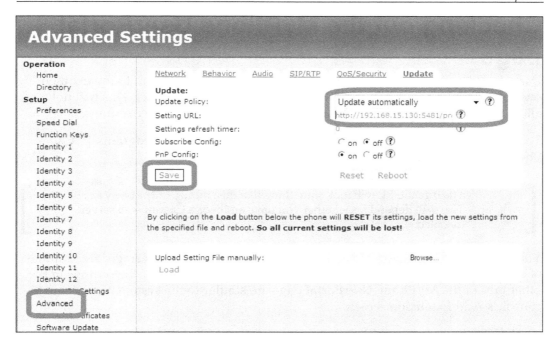

Let's do a couple of quick tests to make sure that the phone works. We can always dial 999 to make sure we have connectivity to the phone system from the phone. We can check the extension status screen in the 3CX administrator console. The light for extension **102** should be green and display **Registered (Idle)**. Now that we have two extensions registered, we can call extension 101 and test that we can call between extensions.

If all the tests pass, then we have successfully provisioned the Snom phone. Now that we have some extensions connected, let's take a more in-depth look at some extension features and settings.

Connecting other phones

There are many other phones that work with the 3CX Phone System. A list of phones that work with 3CX and a detailed instruction on how to provision them can be found at `http://www.3cx.com/sip-phones/index.html`.

Interoperability notes on phones can be found at `http://wiki.3cx.com/phone-configuration/firmwares-tested`.

Checking out the MyPhone UserPortal page

Users have their own **MyPhone UserPortal** page that allows them to review their own phone settings, change some settings, and listen to voicemail. This will not allow them to modify general phone system settings or other user settings.

To log in to your **MyPhone UserPortal** page, go to **Start | All Programs | 3CX Phone System | MyPhone UserPortal**.

 The direct URL will look something like this: `http://3cxServer:5000`. This will depend on whether you have an IIS or a Cassini web server installed.

You will use the **Extension No** and **PIN** to log in. The **Home** screen gives a summary of some important information related to this extension. Many of the settings in the other tabs of the **MyPhone UserPortal** page are similar to the system management console's **Add Extension** screen.

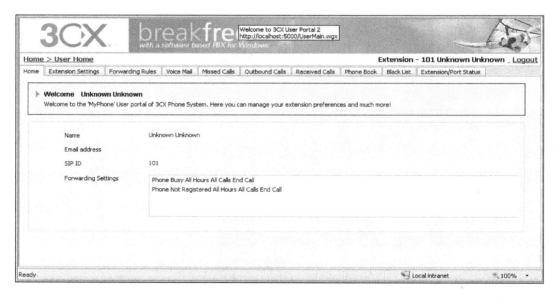

There is one exception though, that is here we have the **Voice Mail** tab.

Voicemail

The **Voice Mail** tab will allow the extension user to listen to and manage voicemail directly from a web page. Messages can be played, deleted, marked as heard, or marked as new.

 Think of the possibilities: you could forward the ports necessary to view this page from the Web, and then your users have access to voicemails wherever they are.

Extension groups

Extensions can be grouped for management and viewing purposes. By default, a new extension is not put into any group; also, extensions can only be a part of one group. On the 3CX management console, **MANAGEMENT** and **UNPOWER USER** are extension groups. Clicking on the **UNPOWER USER** extension group will bring you to the **Edit Extension Group** page, as shown in the following screenshot. The **Edit Extension Group** page allows you to add extensions to this group. As mentioned earlier, if you add an extension to this group, that was a part of another group, it will lose its membership in the other group.

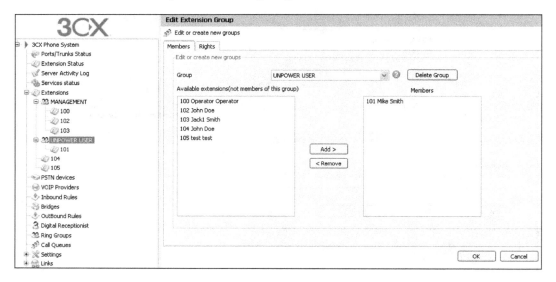

If we click on the **Rights** tab, we can see and edit the rights for this group. On this tab, you can select which users are administrators of this group and who can perform operations on calls in this group by clicking on **Select Extensions**:

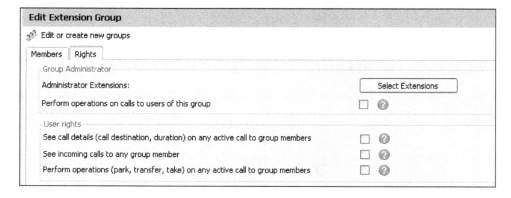

Editing multiple extensions at a time

One last feature related to groups is the ability to select multiple extensions and edit them as a group. To do this, click on **Extensions** on the left side of the **3CX Management Console** as shown in the following screenshot. Now hold the *Ctrl* key while you click on extensions to select multiple extensions. When you have selected the extensions you want to edit, click on **Edit Extension** to edit this group:

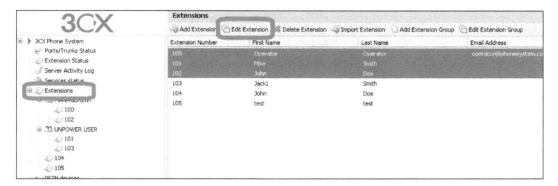

You will notice on the **Edit Extension** page that you are editing three extensions at once. Settings that are specific to one extension or which are not editable are grayed out and locked. Settings that you can change for this group, such as **Enable Voice mail** or **Play Caller ID**, are not locked. Make the changes you want and then click **Apply** as shown in the next screenshot:

 Multiple extension editing allows you to edit provisioning BLF assignments but, according to my tests, you cannot edit these even though they are not locked. The ability to change BLF assignments by groups would be nice but it doesn't seem to work.

Groups are a new feature in version 7 and the usability is somewhat limited but, my guess is, with subsequent releases more features will be added to groups.

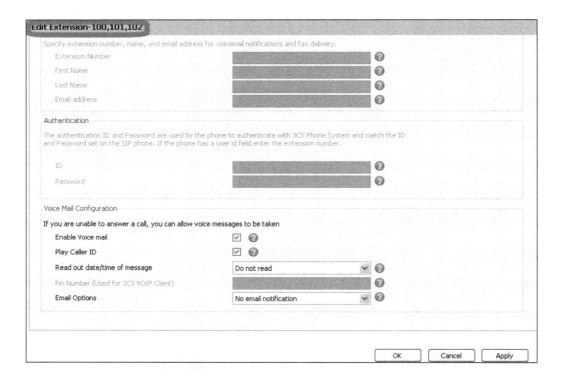

Summary

We've now tested that we can connect to our 3CX Phone System from another computer. We have also set up an extension in 3CX, connected a softphone and phone handset as extensions, and tested to make sure they work correctly. We also got a chance to look at how the MyPhone portal and extension groups work.

We now have a 3CX Phone System installed and extensions configured so that it is largely useable to make calls internally.

In the next chapter, we'll take a look at Digital Receptionist, Hunt Groups, Call Queues, and intercom groups (dial by name), all of which help direct internal and external callers to the right extension.

4
Call Control: Ring Groups, Auto-attendants, and Call Queues

Now that you have made it to Chapter 4, you should have your extensions all set up, or at least a couple of them. If you are unable to call extension-to-extension at this point, then you will need to go back to Chapter 3 and get that working before going any further.

Once your extensions are working, we can begin exploring call routing also called as call control. When someone calls from the outside world, what do you want to do with the call? How do you want your calls to get to an extension? Unless you want your calls to go directly to an extension, you will need to configure one or more of the following features:

- Ring groups (also called Hunt groups in other PBX systems)
- Digital Receptionists or Auto-attendants
- Call by name (also called Dial by Name in some PBX systems)
- Call queues

Let's get started!

Ring groups

Ring groups are designed to direct calls to a group of extensions so that a person can answer the call. An incoming call will ring at several extensions at once, and the one who picks up the phone gets control of that call. At that point, he/she can transfer the call, send it to voicemail, or hang up.

Ring groups are my preferred call routing method. Does anyone really like those automated greetings? I don't. We will of course, set those up because they do have some great uses. However, if you like your customers to get a real live voice when they call, you have two choices—either direct the call to an extension or use a ring group and have a few phones ring at once. To create a ring group, we will use the 3CX web interface. There are several ways to do this.

From the top toolbar menu, click **Add | Ring Group**. In the following screenshot, I chose **Add | Ring Group**:

The following screenshot shows another way of adding a ring group using the **Ring Groups** section in the navigation pane on the left-hand side. Then click on the **Add Ring Group** button on the toolbar:

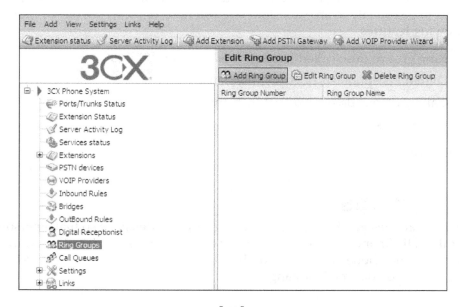

Once we click **Add Ring Group**, 3CX will automatically create a **Virtual machine number** for this ring group as shown in the next screenshot. This helps the system keep track of calls and where they are. This number can be changed to any unused number that you like. As a reseller, I like to keep them the same from client to client. This creates some standardization among all the systems.

Now it's time to give the ring group a **Name**. Here I use **MainRingGroup** as it lets me know that when a call comes in, it should go to the Main Ring Group. After you create the first one, you can make more such as SalesRingGroup, SupportRingGroup, and so on.

We now have three choices for the **Ring Strategy**:

- **Prioritized Hunt**: Starts hunting for a member from the top of the **Ring Group Members** list and works down until someone picks up the phone or goes to the **Destination if no answer** section.
- **Ring All**: If all the phones in the **Ring Group Members** section ring at the same time then the first person to pick up gets the call.
- **Paging**: This is a paid feature that will open the speakerphone on **Ring Group Members**.

Now you will need to select your **Ring Time (Seconds)** to determine how long you want the phones to ring before giving up. The default ring time is **20** seconds, which all my clients agree is too long. I'd recommend 10-15 seconds, but remember, if no one picks up the phone, then the caller goes to the next step, such as a Digital Receptionist. If the next step also makes the caller wait another 10-20 seconds, he/she may just hang up. You also need to be sure that you do not exceed the phone company's timeout of diverting calls to their voicemail (which could be turned off) or returning a busy signal.

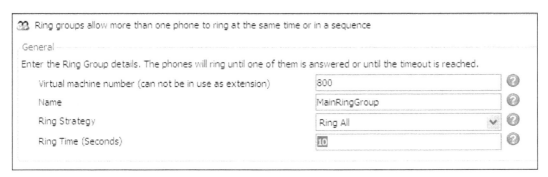

Adding ring group members

Ring Group Members are the extensions that you would like the system to call or page in a ring group. If you select the **Prioritized Hunt** strategy, it will hunt from the top and go down the list. **Ring All** and **Paging** will get everyone at once. The listbox on the left will show you a list of available extensions. Select the ones you want and click the **Add** button. If you are using **Prioritized Hunt**, you can change the order of the hunt by using the **Up** and **Down** buttons.

Destination if no answer

The last setting as shown in the next screenshot illustrates what to do when no one answers the call. The options are as follows:

- **End Call**: Just drop the call, no chance for the caller to talk to someone.

- **Connect to Extension**: Ring the extension of your choice.

- **Connect to Queue / Ring Group**: This sends the caller to a call queue (discussed later in the *Call queues* section)) or to another ring group. A second ring group could be created for stage two that calls the same group plus additional extensions.

- **Connect to Digital Receptionist**: As a person didn't pick up the call, we can now send it to an automated greeting/menu system.

- **Voicemail box for Extension**: As the caller has already heard phones ringing, you may just want to put him/her straight to someone's voicemail.

- **Forward to Outside Number**: If you have had all the phones in the building ringing and no one has picked up, then you might want to send the caller to a different phone outside of your PBX system. Just make sure that you enter the correct phone number and any area codes that may be required. This will use another simultaneous call license and another phone line. If you have one line only, then this is not the option you can use. See Chapter 5, *Trunks – Connecting to the Outside World*, for the line options.

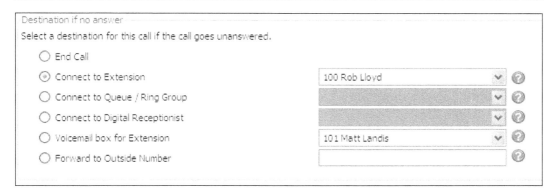

Digital Receptionist setup

A **Digital Receptionist** (**DR**) is not a voicemail box; it's an automated greeting with a menu of choices to choose from. A DR will answer the phone for you if no one is available to answer the phone (directly to an extension or hunt group) or if it is after office hours.

You need to set up a DR unless you want all incoming calls to go to someone's voicemail. You will also need it if you want to present the caller with a menu of options. Let's see how to create a DR.

Recording a menu prompt

The first thing you need to do in order to create a DR is record a greeting. There are a couple of ways to do this. However, first let's create the greeting script. In this greeting, you will be defining your phone menu; that is, you will be directing calls to extensions, hunts, agent groups, and the dial by name directory. Following is an example:

> *Thank you for calling. If you know your party's extension, you may dial it at any time. Or else, please listen to the following options:*
>
> *For Rob, dial 1*
>
> *For the sales group, dial 2*
>
> *For Zachary, dial 4*
>
> *Solicitors, please dial 8*
>
> *For a dial by name directory, dial 9*

I suggest having it written down. This makes it easier to record and also gives the person setting up the DR in 3CX a copy of the menu map.

Now that you know what you want your callers to hear when they call, it's time to get it recorded so that we can import it into 3CX. You have a couple of options for recording the greeting script. It doesn't matter which option you use or how you obtain this greeting file, as long as the end format is correct. You can hire a professional announcer, put it to music, and obtain the file from him/her. You can record it using any audio software you like such as Windows Sound Recorder, or any audio recording software. The file needs to be a `.wav` or an `.mp3` file saved in PCM, 8KHz, 16 bit, Mono format.

If you have Windows Sound Recorder only, I'd suggest that you try out Audacity. **Audacity** is an open source audio file program available at `http://audacity.sourceforge.net/`. Audacity gives you a lot more power such as controlling volume, combining several audio tracks (a music track to go with the announcer), using special effects, and many other cool audio tools. I'm not an expert in it but the basics are easy to do. First, hit the Audacity website and download it, then install it using the defaults. Now let's launch Audacity and set it up to use the correct file format, which will save us any issues later. Start by clicking **Edit | Preferences**. On the **Quality** tab, select the **Default Sample Rate** as **8000 Hz**. Then change the **Default Sample Format** to **16-bit** as shown in the following screenshot:

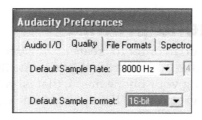

Now, on the **File Formats** tab, select **WAV (Microsoft 16 bit PCM)** from the drop-down list and click **OK**:

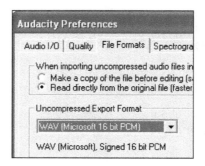

Now that those settings are saved, you can record your greeting without having to change any formats. Now it's time to record your greeting.

Click on the red **Record** button as shown in the following screenshot. It will now use your PC's microphone to record the announcer's voice and when the recording is done, click on the **Stop** button. Press **Play** to hear it, and if you don't like it, start over again:

If you like the way your greeting sounds, then you will need to save it. Click **File | Export As WAV...** or **Export As MP3....** Save it to a location that you remember (for example, c:\3CX prompts is a good place) with a descriptive filename. While you are recording this greeting, you might as well record a few more if you have plans for creating multiple DRs:

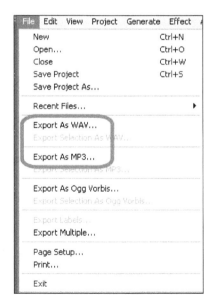

Creating the Digital Receptionist

With your greeting script in hand, it's time to create your first DR. In the navigation pane on the left side, click **Digital Receptionist**, then click **Add Digital Receptionist** as shown in the following screenshot:

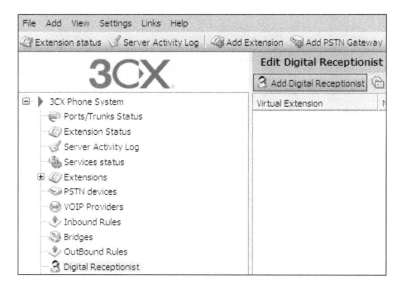

Or on the top menu toolbar, click **Add | Digital Receptionist**:

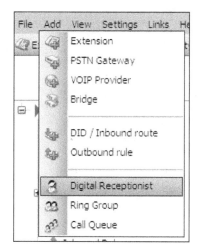

Just like your ring group, the DR gets a **Virtual extension number** by default, Feel free to change it or stick with it. Give it a **Name**, (I like to use the same name as the audio greeting filename.) Now, click **Browse...** and then **Add**. Browse to your `c:\3CX prompts` directory and select your `.wav` or `.mp3` file as shown in the following screenshot:

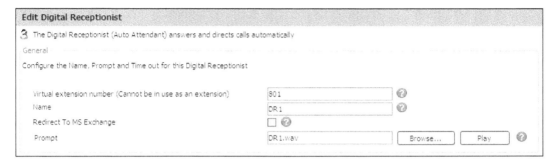

Next, we need to create the menu system as shown in the following screenshot. We have lots of options available. You can connect to an extension or ring group, transfer directly to someone's voicemail, end the call (my solicitors' option), or start the call by name feature (discussed in the *Call by name setup* section). At any time during playback, callers can dial the extension number; they don't have to hear all the options. I usually explain this in the DR recorded greeting.

It's a good idea to set the timeout to connect to an extension or transfer it to voicemail, just in case they do not have a touch-tone **Dual-tone Multi-frequency (DTMF)** phone.

Click **OK** at the bottom to save the menu, and you're done:

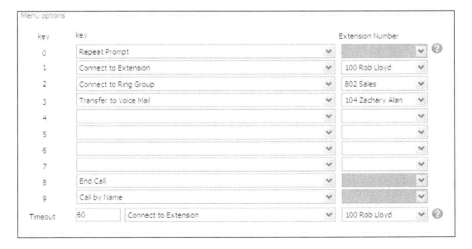

Previously, when we made a ring group, we didn't have the DR done. At that time, we weren't able to choose this option if no one was available to pick up the phone.

Now, you can go back to the **Ring Group** and, in the **Destination if no answer** section, select the **Connect to Digital Receptionist** option that you just created. It gives callers a chance to talk to a human first, and then if it goes unanswered, they will get the automated greeting. This is my preferred method which works very well for a home or a small business where there still might be a receptionist.

Call by name setup

This is a great option to have if you have lots of people in your company and you don't want to list them all in your DR. If your greeting script doesn't have room for every extension to be listed in it and the caller knows the person's name, this will let him/her look it up by dialing the name. There are three requirements that need to be met in order for the call by name function to work:

- Users must have a self-identification message. Without this message, they will not be available for the feature.
- The user's last name must be a-z or 2-9.
- The call by name feature must be enabled in the Digital Receptionist menu, as shown in the previous screenshot, by having the caller select option 9 (for example).

Here is how the system works:

When you set up the extensions, you will need to tell all the users to record their self-identification number. This is done by accessing their voicemail (999 by default). Now go to the Options Menu by dialing 9. Then, dial 5. If you do not have one already recorded, it will prompt you to do so. If you've already recorded it and would like to change it, dial 0. If you want to delete it to remove them from this feature, dial 1.

Callers will now have to press the appropriate numbers for the last name. They must dial a minimum of three digits. To search for "Lloyd,", they would dial 556. If the person's last name is only two characters, they can dial 0 to fill in the third digit.

Once three digits are dialed, the system will search for a match. If it can't find a match, callers will hear *Extension not found*. If there is a match, they will hear *Please hold while I'm calling to Rob Lloyd* (my self-identification message).

If there is more than one match, the system will wait two seconds for an additional digit to help separate the matches. It will repeat this process until there is no longer a match, and then the caller will hear *Extension not found*.

If callers let the two-second timeout elapse or they press #, they will hear a menu of matches such as *To call Ed Jones press 0, to call Sam Jonson press 1, or to exit press pound.*

Call queues

Another paid feature of 3CX is **call queues**. A call queue is a holding area for callers to wait until someone in the queue group is available to get the next caller. I'm sure everyone has been stuck in a queue for support.

Here's how it works. Callers have a support issue they need help with. They call your company and get connected to the Digital Receptionist. They dial 3 for support. If you have two people doing all the support calls for this product, then when callers dial 3, they get put in the queue. They will hear some music until one of the available support staff is off the phone. When the staff hangs up, callers will be transferred to their extension.

Let's create a queue! We can create a queue by using the usual method: Click **Add | Call Queue** on the menu toolbar as shown in the following screenshot:

Or, in the navigation pane on the left-hand side, click **Call Queues**, then on the right-hand side, click **Add Queue** as shown in the next screenshot:

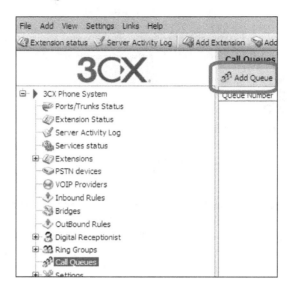

Just like the DR and ring groups, we have a queue **Virtual Extension Number,** too. Leave the default or change it to an unused number. Give it a descriptive **Name**, as shown in the following screenshot. I called mine **SupportQueue**.

Now you have the option for the **Ring timeout(seconds)** which will ring the support agent's phone for **30** seconds before being placed back in the queue.

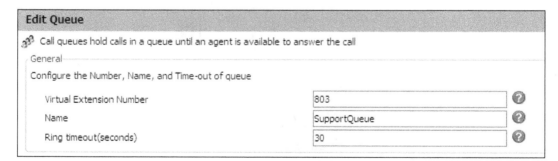

In the next section, select your **Call Queue Agents** just like you did with the ring group, as shown in the following screenshot. Use the **Up** and **Down** buttons to change the priority of the extensions. The priority is used to determine who will get the call if there is more than one agent available at the same time.

The queue agents **must** log in to the call queue by using the 3CX VoIP Client or by using dial codes on the phone. The VoIP Client and dial codes are discussed in a later chapter:

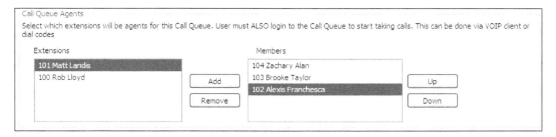

The next section is **Destination if no answer**. Here, we can see the various options available if no one picks up the phone, no one is logged into the queue, or the caller presses the * button. You should always provide some kind of fallback for the caller to reach someone or to get out of the queue:

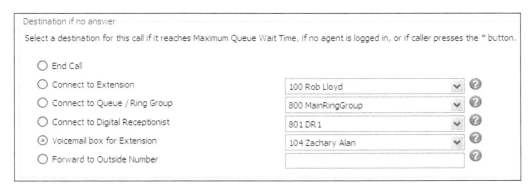

The next set of **Other Options** as shown in the next screenshot are the customization options for this particular call queue. They are as follows:

- **Enable intro prompt**: This option gives the caller an introduction prompt—*Thank you for calling. You are now in the queue. Please enjoy the music.* The audio file needs to follow the same format as our Digital Receptionist.

- **Announce Queue position to caller**: It's nice to know how deep into the queue you are. If I get into a queue with 20 people in front of me, I would probably hang up and try again later. If I hear I'm second in line, I'll gladly wait.

- **Announcement Interval (seconds)**: This timer is used to update callers on their queue status. It's nice to hear that I'm getting closer to a support person.

- **Music on hold**: If this is a sales queue, I might just record some advertisement material such as *Thank you for waiting. This month we have our new product on sale. Ask your sales representative for details*. We will see this in detail in Chapter 6 in the *For the iTunes user* section.

- **Maximum Queue Wait Time (seconds)**: This timer gives the caller a maximum time before we get to the Destination if no answer mode.

Click **OK** when you are done to save your changes:

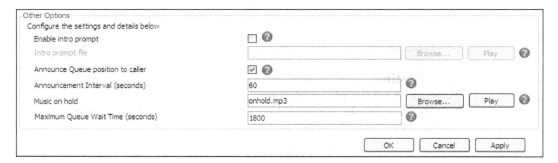

Summary

Controlling calls is an essential part of any phone system. Without call control, every incoming phone call would be sent directly to someone's extension. In this chapter, we looked at various ways to control and handle calls in the 3CX Phone System. We learned how to set up a ring group, DR, and one of those dreaded call queues.

We also learned how to set up the call by name option in the DR and learned how it works. Don't forget that this needs to be turned on in the Digital Receptionist menu, and the individual extensions need to be set up. In the Digital Receptionist greeting, give the caller some instructions on how it works.

Go set them up and test them out. When testing, make sure every option works. You do not want a caller to be stuck somewhere unable to do anything except hang up.

In the next chapter, we will discuss how to get those incoming calls connected to 3CX using a trunk.

5
Trunks—Connecting to the Outside World

We now have a working phone system with extensions, voicemail, digital receptionists, call queues, and several other features. This is great! However, what if we want to call home or have a customer call us? For that, we need to connect 3CX outside our internal network. This connection is called a **trunk**. In this chapter, we will cover the following:

- PSTN trunks
- SIP trunks
- Introduction to dial plans (full coverage is in Chapter 6, *Configuration*)
- Hardware needed for analog lines

PSTN trunks

A **Public Switched Telephone Network (PSTN)** trunk is an old-fashioned analog **Basic Rate Interface (BRI)** ISDN or **Primary Rate Interface (PRI)** phone line. 3CX can use any of these with the correct analog to SIP gateway. Usually, these come into your home or business through a pair of copper lines. Depending on where you live, this may be the only means of connecting 3CX and communicating outside of your network.

One of the advantages of a PSTN line is reliability and great call quality. Unless the wires break, you will almost always have phone service. However, what about call quality? After all, many people would like to have a comparison between VoIP and PSTN.

Analog hardware for BRI ISDN and PRIs will be discussed in greater detail in Chapter 9. For using an analog PSTN line, you will need an FXO gateway. There are many external ones available. Until Sangoma introduced a new line at the end of 2008, there had not been any gateway that worked inside a Windows PC with 3CX. There are many manufacturers of analog gateways, such as Linksys, AudioCodes, Patton Electronics, Grandstream, and Sangoma. What these FXO gateways do is convert the analog phone line into IP signaling. Then, the IP signaling gets passed over your network to the 3CX server and your phones.

My personal preference is Patton Electronics. They are probably the most expensive FXOs out there but, in this case, you get what you pay for. I have tried all of them and they all work. Some have issues with echo, which can be hard to get rid of without support or lots of trial and error, whereas some cannot support high demands (40 calls/hour) without needing to be reset every day. So, if you are just testing, get a low-end one. For a high-demand business, my preference is Patton. Not only do they make great products, but their support is top notch, too. We will configure a Patton SmartNode SN4114 later in this chapter.

SIP trunks

What is a SIP trunk? A **SIP trunk** is a call that is routed by IP over the Internet through an **Internet Telephony Service Provider (ITSP)**.

For enterprises wanting to make full use of their installed IP PBXs and communicate over IP not only within the enterprise, but also outside the enterprise, a SIP trunk provided by an ITSP that connects to the traditional PSTN network is the solution. Unlike traditional telephony, where bundles of physical wires were once delivered from the service provider to a business, a SIP trunk allows a company to replace traditional fixed PSTN lines with PSTN connectivity via a SIP trunking service provider on the Internet.

SIP trunks can offer significant cost savings for enterprises, eliminating the need for local PSTN gateways, costly ISDN BRIs or PRIs. The following figure is an example of how our phone system operates:

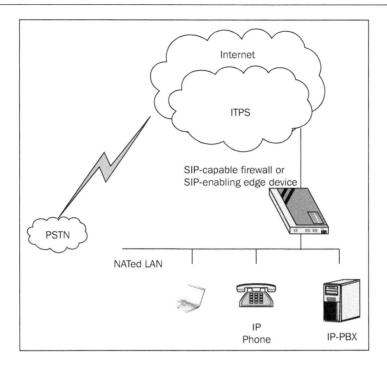

You can see that we have a local area network containing our desktops, servers, phones, and our 3CX Phone System. To reach the outside world using a SIP trunk, we have to go through our firewall or router. Depending on your network, you could be using a private IP address (10.x.x.x, 172.16.x.x, or 192.168.x.x), which is not allowed on the public Internet, so it has to get translated to the public IP address. This translation process is called **Network Address Translation (NAT)**.

Once we get outside the local network, we are in the public realm. Our ITSP uses the Internet to get our phone call to/from the various carriers' PSTN (analog) lines where our phone call is connected/terminated.

In Chapter 10, we will cover NAT, ports, and some router tips and tricks to connect to the ITSP and also to remotely connect to our 3CX Phone System.

There are three components necessary to successfully deploy SIP trunks:

- A PBX with a SIP-enabled trunk side
- An enterprise edge device understanding SIP
- An Internet Telephony or SIP trunking service provider

The PBX

In most cases, the PBX is an IP-based PBX, communicating with all end points over IP. However, it may just as well be a traditional digital or analog PBX. The sole requirement that has to be available is an interface for SIP trunking connectivity.

The enterprise border element

The PBX on the LAN connects to the ITSP via the enterprise border element. The enterprise edge component can either be a firewall with complete support for SIP or an edge device connected to the firewall handling the traversal of the SIP traffic.

The ITSP

On the Internet, the ITSP provides connectivity to the PSTN for communication with mobile and fixed phones.

Choosing a VoIP carrier—more than just price

Two of the most important features to look for when choosing a VoIP carrier are support and call quality. Usually, once you set up and everything is working, you won't need support. I always tell clients that there is no "boxed" solution that I can sell; every installation is a little different. Internet connections are all different even with the same provider. If you have a rock-solid T1 or something better, then this shouldn't be a problem. DSL seems different from building to building, even in the same area.

So how do you test support before giving them your credit card? Call them! Try calling support at the worst times, such as Monday afternoons when everyone is back to work and online. Also, try calling after business hours. See how long it takes to connect to a live person and if you can understand them once you speak to them. Find out where their support is located. Try talking to them and tell them you are thinking about signing up with their service and ask them for help. If they go out of their way before they have your money, chances are they will be good to work with later on. Some carriers only offer chat or e-mail support in favor of lower prices. While this may work fine for your business, it certainly won't work for the ones who need answers right away.

I know I seem to be stressing a lot on support, but it's for good reason. If your business depends on phone service, and your phone service goes down, then you need answers! I pay more for a product if the support is worth it. Part of this is your **Return On Investment** (**ROI**). For example, if you have three lawyers billing at $200/hour and they need phones to work, that's $600/hour of lost time. Does the extra $50 or $100 upfront cover that? Now, back to the topic at hand.

Once you have connected 3CX to the carrier, how is the call quality? If it sounds like a bad cell phone, you probably don't want it, unless the price is so cheap that you can live with the low quality. Certain carriers even change the way your call gets routed through the Internet based on the lowest cost for the particular call. They don't care about quality as long as you get that connection and they make money on it.

Concurrent calls with an ITSP are a feature that you may want to look for when choosing an ITSP. Some accounts are a one-to-one ratio of lines per call. If you want to have five people on the phone at the same time (inbound or outbound), you would need to pay for five lines. This is similar to a PSTN line. You may get some savings here over a PSTN, but that depends on what is available in your area.

Some ITSPs have concurrent calls where you can use more than one line per call. Not many carriers have this feature but, for a small business, this can be a great money-saving feature to look for. I use a couple of different carriers that have this feature.

One carrier that I use lets you have three concurrent calls simultaneously on the same line. If you need more than three calls, you're a higher-use customer, and they want you to buy several lines.

VoIP IP signaling uses special algorithms to compress your voice into IP packets. This compression uses a codec. There are several available, but the most common one is **G.711U-law** or **A-law**. This uses about 80Kpbs of upload and download bandwidth. Another popular codec is **G.729**, which uses about 36Kpbs. So, for the same bandwidth, you can have twice the number of calls using G.729 than G.711. You will need to check with your ITSP and see what codec they support.

Another carrier I use is based purely on how much Internet bandwidth one has. If you have 1MBps of upload speed (usually the slowest part of your Internet connection), you can support about 10 simultaneous or concurrent calls using G.711. You then pay for the minutes you use. This works very well for a small office as your monthly bill is very low, and you don't have to maintain a bunch of lines that don't get used.

Cable Internet providers are also offering VoIP services to your home or business. These are usually single-use lines, but they terminate at your office with an FXS plug. To integrate this with 3CX, you will need an FXO just like it's a PSTN line. It's the same setup, but you get the advantage of a VoIP line.

Another great benefit of a SIP trunk is expandability. You can easily start out with one line that can usually be completed in one day. As you grow, you can add more, usually in minutes as you already have the plan set up. Time to consolidate lines? You can even drop them later on without having contracts (most of the time). Try doing that with the local phone company! Call for a new business and it can take 1-2 weeks to get set up, plus contracts to worry about. No wonder they are jumping on the VoIP bandwagon.

Disaster recovery

What do you do when your Internet goes down? Some of you might be saying, "Ha! It never goes down." In my experience, it will eventually, and at the worst time. So, what do you do? Go home for the day or plan for a backup? Most VoIP carriers provide some kind of disaster recovery option. They try to send you a call and, when they don't get a connection to your 3CX box, they re-route the call to another phone number. This could be a PSTN line or even a cell phone. It can be a free feature or there can be a small monthly fee on the account. It's worth having, especially if you rely on phones.

Okay, so that covers inbound disaster recovery. What about outbound? Yes, just about everyone has a cell phone these days. If that isn't enough, I'd suggest you invest in a pay-per-use PSTN line. This keeps the monthly cost very low, but it's there when you need it. Whether it's an emergency pizza order for that Friday afternoon party or a true emergency, when someone panics and dials 911, you want that call to go out.

Speaking of emergency numbers, make sure you have your carrier register that phone number to your local address. Let's say you are in New York and you have a Californian phone number to give you some local presence in that part of the country. Your co-worker grabs his chest and falls down and someone dials 911 from the closest phone he/she sees. Emergency services sees your Californian number and contacts California for help for your New York office. That's not what you want when your co-worker is clutching his/her chest, even though it was just heartburn from that pepperoni pizza.

Mixing VoIP and PSTN

Some of my clients even mix VoIP and PSTN together. Why would you mix? Local calls and inbound calls use the PSTN lines for the best call quality (and do not use any VoIP minutes if they have to pay for those). Long distance calls use the cheaper rate VoIP line. Another scenario is using PSTN lines for all your incoming and outgoing calls and using VoIP to talk to your other offices. Your own office can deal with a lower call quality, and management will appreciate the lower cost.

These types of setups can be controlled using a dial plan discussed in Chapter 6, *Configuration*.

Connecting 3CX to your trunk

Let's cover the setup for connecting 3CX to a PSTN line using an analog gateway (Patton SN4114) and then connecting 3CX to a SIP/ITSP line.

The first thing you need to know is that every line or port in 3CX is assigned its very own number, just like the Ring Groups, Digital Receptionists, and Call Queues have their own account number assigned. This makes it easier to route calls using a number.

Let's get started with creating an analog trunk:

The first thing we need to do is start the PSTN Gateway wizard. We can do this in three different ways. It does not matter which method you use as they all start the wizard the same way.

1. The first way is to click **Add** and then **PSTN Gateway**:

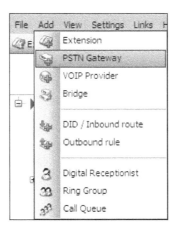

2. The second way is to click **Add PSTN Gateway** on the main 3CX toolbar:

3. The third method is to use the navigation pane on the left-hand side; click **PSTN devices**, and then click **Add Gateway** on the right-hand side:

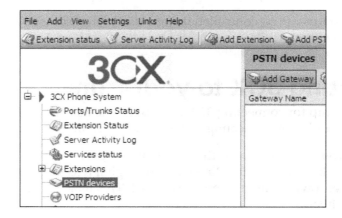

Now that we have started the PSTN Gateway wizard, we can run through the steps. First, come up with a name; I'd suggest something meaningful. I like to use the model number and something else after it, like an *A* or a *1*. Using this method lets you expand easily and keep the naming conventions the same for all devices. As I'm using a Patton SN4114 gateway, I chose the name **PattonSN4114A**. If I need to add another gateway, then I can use the same name and use a *B* at the end. Using a label maker, I also label the Patton itself with the name, IP address, and (depending on the environment) maybe even the username and password.

The next step is to pick which supported gateway you have, and you can see there are a lot of choices. If you are using a Patton, you need to know which firmware the device has on it. Version 4 and version 5 firmware need slightly different configuration files. 3CX is smart enough to know the difference when it generates the configuration file. This configuration file is going to be saved and uploaded to the Patton to instantly configure the gateway. Pretty cool! Otherwise, you have a choice of configuring the gateway using a command line connected to a serial port cable on your computer and the Patton, or using their clunky web interface—neither are user-friendly.

Just to save some space, I have cut out a link of some of the supported gateways in the following screenshot. For a complete list of supported gateways, follow the link.

There are more gateways that work with 3CX that are listed on their website, but they don't work with the gateway wizard. If you ever need support from a vendor or 3CX directly, I'd suggest that you use one that is listed in the wizard.

I know my gateway has version 4 of the firmware because it says so on the box. If you don't have the box it came with, or you just aren't sure, you will have to go into the web interface or the console command line and obtain the firmware version. Here, I'm selecting **Patton SN-4114 4-port FXO (Firmware R4.x)**.

After you select your gateway, click **Next**:

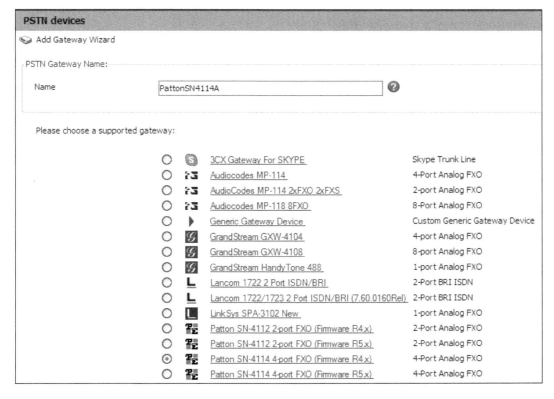

Other supported gateways have been cut for space reasons.

On the next screen, as shown in the following screenshot, you will have to select a few options. The first one is **Tone Set Selection** to select which country this gateway is going to be installed in.

The next section is for **Incoming Caller ID** info. If you have caller ID, you will want to select **Collect CallerID information**. This will delay the gateway from picking up the incoming call immediately. It will wait another ring (depending on which country and your phone company information) before picking up the call. Once it waits a second or two, it should have all the caller ID information, answer the phone line, and pass it to 3CX for processing (what to do with the call).

Our next section is to **Remove Announcements** that the phone company passes on. I don't like these, so I have them removed.

If we are going to use this gateway for outbound calls, 3CX needs to know how to do it. We can hunt (look for) a free line on the gateway. It's based on port numbers, so if you want it to start looking at port 1000 for a free line, and then 1001, choose **Hunting (Ascending)**. If you want it to start at the end (1004) and then go to 1003, choose **Descending**. If you have roll-over lines, you may want choose **Descending**, but it doesn't really matter to the Patton.

Our last setting covers how long you want to wait to collect the caller ID information. Depending on your phone carrier, you may have to adjust this setting. If you find that you are not getting all the information, you will have to edit this and adjust accordingly.

Now that this screen is all filled in, we can click **Next**:

Our next wizard screen is device specific. We start off by giving it a **Gateway Hostname or IP** address. Unless you have your own DNS server or are using WINS or host files, you will want to use an IP address. Even if you have a DNS server, I'd still suggest you use an IP address. You certainly don't want to lose your connection to the gateway if your DNS server is down.

Now, we need to specify which **Gateway Port** to use. Unless you have a reason to use something different, stick with the well-known default SIP port **5060**.

The next setting is how many ports we are going to use for this gateway. The default is the maximum number of ports that the device has available. Even if you don't use them all, it will be easy to upgrade if necessary. When you're done with this screen, click **Next**:

Now we get to create the port numbers, names, passwords, and some rules for call processing. We can see here that we have a **Virtual extension** number, **Authentication ID**, and **Authentication Password**. If you want to change these to something different, now is the time to do it. The best name to use here is the actual line number.

As we are using analog single call lines, we need to leave the **Channels** section to **1**.

The only real thing we may want to change is the **Inbound Route Day** and **Inbound Route Night**. This tells 3CX what to do with an incoming call during the day and what to do at night. (Day/night settings are discussed elsewhere in this book.) During the day, I want it to go to the hunt group that we defined in the previous chapter. At night when no one is around to answer the call, I set it to go directly to the Digital Receptionist.

Go ahead and click **Next**:

PSTN devices								

🐌 Create ports

The following ports will be created in the "Create Ports" screen. You can edit the Port identification and authentication settings before they are created. Note that the Port identification is used for identification purposes, and the internal line number is used by 3CX Phone System to address the line connected to the port on the VOIP Gateway. Therefore the Internal Line Number range should be different from the extension number range. You can configure to which extension incoming calls should be routed based on whether they are inside or outside office hours (inbound route).

Remove selected	Virtual extension	Authentication ID	Authentication Password	Channels	Port Identification	Inbound Route Day	Inbound Route Night
☐	10000	10000	10000	1	10000	800	801
☐	10001	10001	10001	1	10001	800	801
☐	10002	10002	10002	1	10002	800	801
☐	10003	10003	10003	1	10003	800	801

< Back   Next >

The next wizard screen is for **Outbound Call Rules**. We will discuss this in greater detail in the next chapter, but let's go over it enough so that you can set up a simple rule.

We start off with a name. This can be anything you like, but I prefer something meaningful. For our example, I want to dial 9 to use the analog line, and only allow extensions 100-102 to use this line. I also only want to be able to dial certain phone numbers. Then, I have to delete the 9 before it goes out to the phone carrier. Let's have a look at each section of this screen:

Calls to numbers starting with (Prefix)

This is where you specify what you want someone to dial before the line is used. You could enter a string of numbers here to use as a "password" to dial out. You don't just let anyone call an international phone number, so set this to a string of numbers to use as your international password. Give the password only to those who need it. Just make sure you change it occasionally in case it slips out.

Calls from extension(s)

Now, you can specify who (by extension number) can use this gateway. Just enter the extension number(s) you want to allow either in a range (100-110), individually (100, 101, 104), or as a mix (100-103, 110). Usually, you will leave this open for everyone to use; otherwise, you will restrict extensions that were allowed to use the gateway, which will have repercussions of forwarding rules to external numbers.

Calls to numbers with a length of

This setting can be left blank if you want all calls to be able to go out on this gateway. In the next screenshot, I specified **3**, **7**, **10**, and **11**. This covers calls to 911, 411, 555-1234, 800-555-1234, and 1-800-555-1234, respectively. You can control what phone numbers go out based on the number of digits that are dialed.

Route and strip options

Since this is our only gateway right now, we will have it route the calls to the Patton gateway. The **Strip Digits** option needs to be set to **1**. This will strip out the "9" that we specified above to dial out with. We can leave the **Prepend** section blank for now.

Don't worry! We will cover these in greater detail with more powerful options in Chapter 6. Go ahead and click **Finish**:

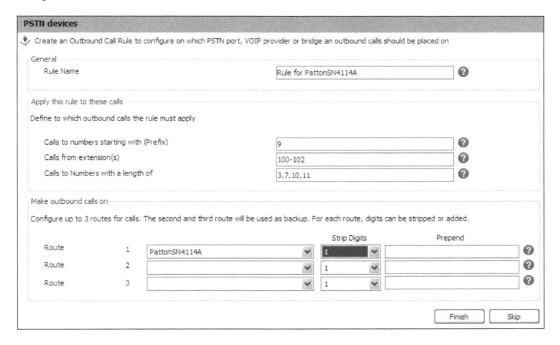

Once you click **Finish**, you will see a gateway wizard summary, as shown in the next screenshot. This shows you that the gateway is created, and it also gives an overview of the settings. Your next step is to get those settings configured on your gateway.

There is a list of links for various supported gateways on the bottom of the summary page with up-to-date instructions. Feel free to visit those links. These links will take you to the 3CX website and explain how to configure that particular gateway. With Patton this is easy; click the **Generate config file** button.

The only other information you need for the configuration file is the **Subnet mask** for the Patton gateway. Enter your network subnet mask in the box. Here, I entered a standard Class C subnet mask. This matches my 192.168.X.X network. Click **OK** when you are done:

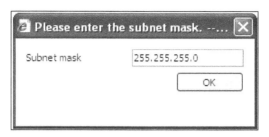

Once you click **OK**, your browser will prompt you to save the file, as shown in the following screenshot. Click **Save**:

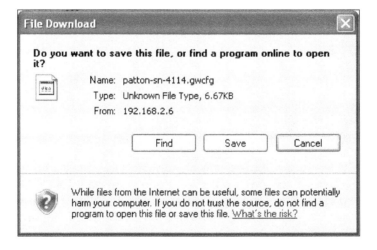

The following screenshot shows a familiar **Save As** Windows screen. I like to put this file in an easy-to-remember location on my hard drive. As I already have a 3CX folder created, I'm going to save the file there. You can change the name of the file if you wish. Click **Save**:

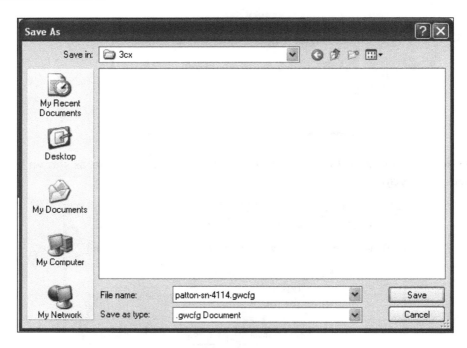

Now that your file is saved, let's take a look at modifying those settings. Open the administration web interface and, on the left-hand side, click **PSTN Devices**. Go ahead and expand this by clicking the **+** sign next to it. Now, you will see our newly created **Patton SN4114A** gateway listed. Click the **+** sign again and expand that gateway.

Next, click the **Patton SN4114A** name, and you will see the right-hand side window pane fill up with five separate tabs.

The first tab is **General**. This is where you can change the gateway IP address, SIP port, and all the account details. If you change anything, you will need a new configuration file. So click the **Generate config file** button at the bottom of the screen. If you forgot to save the file previously, here's your chance to generate and save it again:

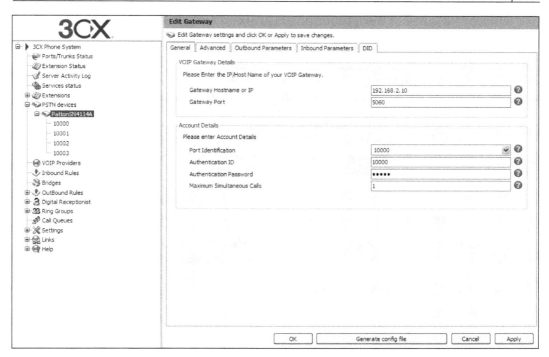

On the **Advanced** tab, we have some **Provider Capabilities**. Leave these settings alone for now as we will discuss these in the *Creating a SIP trunk* section coming up later in this chapter:

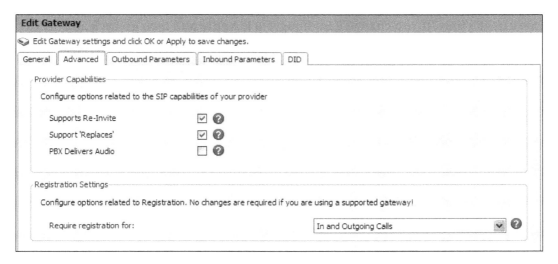

We will leave the rest of the tabs for now. Go ahead and click the **10000** line information in the navigation pane on the left.

These are the settings for that particular phone port (10000). The first group of settings that we can change is the authentication username and password. Remember, this is to register the line with 3CX and not to use the phone line.

The next two sections are about what to do with an inbound call during **Office Hours** and **Outside Office Hours**. I didn't change anything from the gateway wizard but, on this screen, you can see that we selected Ring group **800 MainRingGroup**. This is the Ring group that we configured previously.

We also see similar drop-down boxes for **Outside Office Hours**. As no one will be in the office to answer the phone, I've selected a Digital Receptionist **801 DR1**.

In the section **Other Options**, the **Outbound Caller ID** box is used to enter what you would like to have presented to the outside world as caller ID information. If your phone carrier supports this, you can enter a phone number or a name. If the carrier does not support this, just leave it blank and talk to your carrier as to what you would require to have it assigned as your caller ID.

The **Allow outbound calls on this line** and **Allow incoming calls on this line** checkboxes are used to limit calls in or out. Depending on your environment, you might want to leave one line selected as **no outbound calls**. This will always leave an incoming line for customers to call. Otherwise, unless you have other lines that they can call on, they will get a busy signal.

Maximum simultaneous calls cannot be changed here as analog lines only support one call at a time. If you changed anything, click **Apply** and then go back and generate a new configuration file:

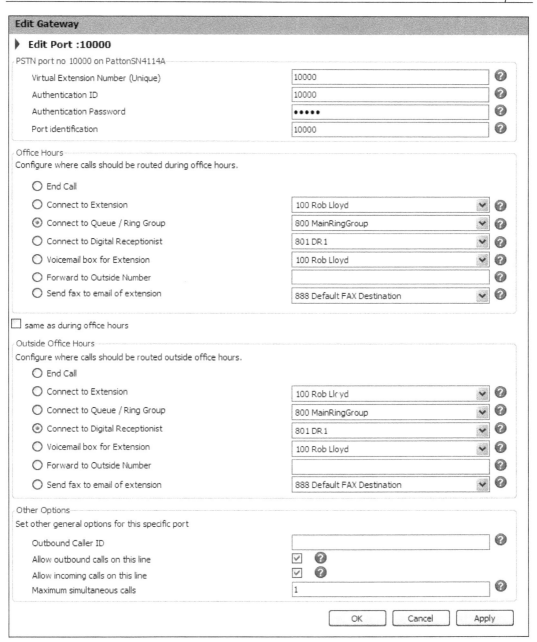

Edit Gateway

▶ **Edit Port :10000**

PSTN port no 10000 on PattonSN4114A

Virtual Extension Number (Unique)	10000	
Authentication ID	10000	
Authentication Password	•••••	
Port identification	10000	

Office Hours
Configure where calls should be routed during office hours.

- ○ End Call
- ○ Connect to Extension — 100 Rob Lloyd
- ◉ Connect to Queue / Ring Group — 800 MainRingGroup
- ○ Connect to Digital Receptionist — 801 DR1
- ○ Voicemail box for Extension — 100 Rob Lloyd
- ○ Forward to Outside Number
- ○ Send fax to email of extension — 888 Default FAX Destination

☐ same as during office hours

Outside Office Hours
Configure where calls should be routed outside office hours.

- ○ End Call
- ○ Connect to Extension — 100 Rob Llryd
- ○ Connect to Queue / Ring Group — 800 MainRingGroup
- ◉ Connect to Digital Receptionist — 801 DR1
- ○ Voicemail box for Extension — 100 Rob Lloyd
- ○ Forward to Outside Number
- ○ Send fax to email of extension — 888 Default FAX Destination

Other Options
Set other general options for this specific port

Outbound Caller ID		
Allow outbound calls on this line	☑	
Allow incoming calls on this line	☑	
Maximum simultaneous calls	1	

OK   Cancel   Apply

[For the most up-to-date information on configuring your gateway, visit the 3CX site: `http://www.3cx.com/voip-gateways/index.html`]

We will go over a summary of it here:

Since nothing was changed, it is now time to configure the Patton device with the config file that we generated from the 3CX template. If you know the IP address of the device, go ahead and open a browser and navigate to that IP address. Mine would be `http://192.168.2.10`. If you do not know the IP address of your device, you will need the SmartNode discovery tool. The easiest place to get this tool is the CD that came with the device. You can also download it from `http://www.3cx.com/downloads/misc/sndiscovery.zip`, or search the Patton website for it.

Go ahead and install the SmartNode discovery tool and run it. You will get a screen that tells you all the SmartNodes on your network with their IP address, MAC address, and firmware version. Double-click on the **SmartNode** to open the web interface in a browser.

The default username is **administrator**, and the password field is left blank.

Click **Import/Export** on the left and **Import Configuration** on the right. Click **Browse** to find the configuration file that we generated. Click **Import** and then **Reload** to restart the gateway with the new configuration.

That's it. We can now get incoming calls and make an outbound call.

Creating a SIP trunk

Now it's time to create a SIP trunk. The best thing to do is use one of the supported VoIP providers. You can use anyone if you get the correct information but, if you have problems, 3CX won't be able to help you out.

Just like setting up the PSTN line, you can get to the VoIP provider wizard in the same manner. I'm going to click **Add VOIP Provider Wizard** on the top menu bar. Pick the method you are most comfortable with.

The first thing you want to do is see which VoIP providers are supported and set up an account on their website. While the 3CX wizard is quick and easy, it cannot create your account with an ITSP. As I'm in the US and Callcentric lines are free (as long as the call is to another Callcentric customer), I am going to use a fake one to get us started. Go ahead and set up the account with the provider of your choice, and get your account details. Once you get the account information, it's time to move on and enter it into 3CX so that you can make calls.

It's time to enter the VoIP provider name that you want to use. Select the VoIP provider that you signed up with and click **Next**:

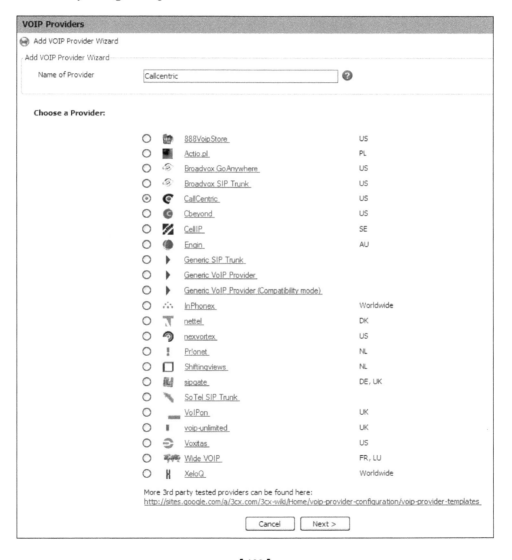

The next screen is for our information only. If we had chosen a generic VoIP provider, we would need to enter the **SIP server hostname or IP**, the **SIP Server port**, and the **Outbound proxy hostname or IP**, and the **Outbound proxy port**. As this is a supported provider, 3CX has this information in the wizard template. Click **Next**:

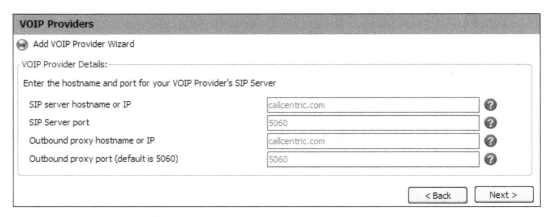

Now in the **Account Details** section, enter the phone number that we were assigned from the provider in the **External Number** field, and also enter the username and password.

The next section is to define how many **Simultaneous Calls** we can have at one time on this account. This is one of those providers that allow multiple concurrent calls, which is great for the small business or home office. Enter the appropriate number and click **Next**:

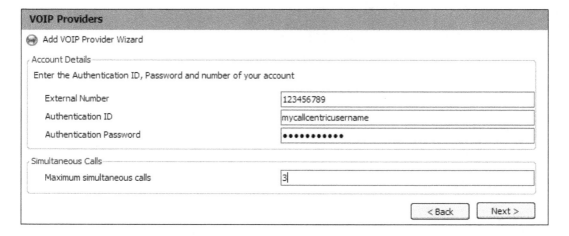

Just like our PSTN call processing, we have the same set of options for our VoIP provider when a call comes in. Here, I wanted the incoming calls to go to the **802 Sales** ring group during office hours. Once everyone has left for the day, I will have it go to the voicemail box of **104 Zachary Alan**. Click **Next**:

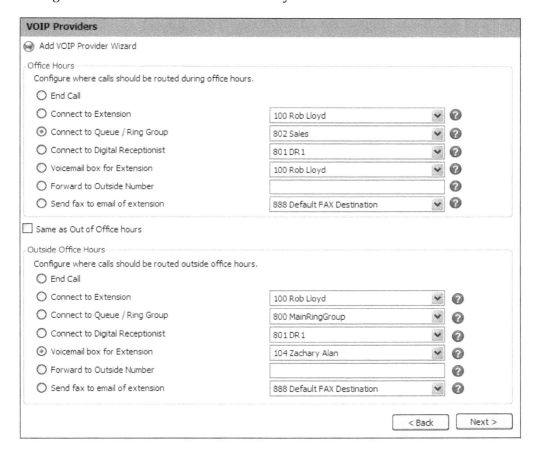

Now, we need an **Outbound Call Rule**. As we already went over most of these options and will do so in greater detail in the next chapter, I'm going to make only one change.

As VoIP calls are less expensive than my PSTN calls, I want to allow everyone to use it. So, "no rules" in the upper section this time.

In the **Make outbound calls on** section, we want everyone to use the **Callcentric** route. We didn't use the **Calls to numbers starting with (Prefix)**, so we don't want to strip any digits. However, we do want to **Prepend** a **1** before every call. Most VoIP providers treat every call as though it is long distance, even if it's across the street. To help avoid having people dial "1" with every call, we can have 3CX do this automatically. Click **Finish**:

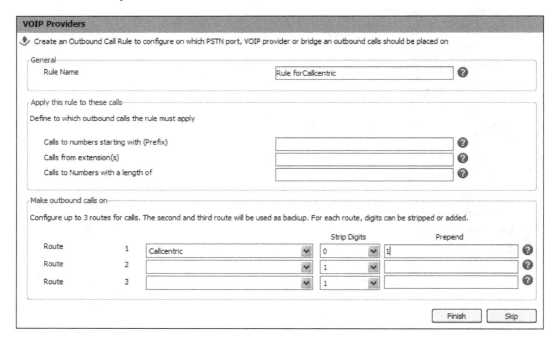

In the navigation pane on the left-hand side, click **VOIP Providers** and expand it by clicking the **+** sign next to it:

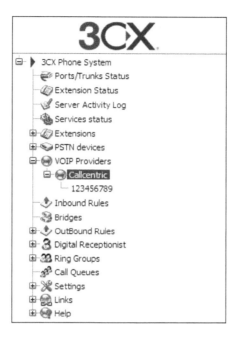

In the right-hand window pane, we can see our information for this provider. If you need to change any of this information (like fixing a typo during the wizard), go ahead and do so here:

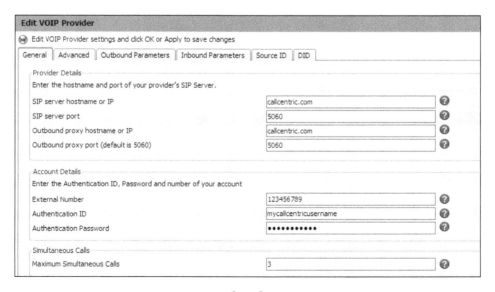

Click on the **Advanced** tab, as shown in the following screenshot. This screen is a little different from the PSTN screen as we now have **Registration Settings**, too. These settings define how often the trunk line re-registers with the VoIP provider. It is a good thing because if the time was indefinite and your Internet went down, 3CX would still try to make calls on this line.

The bottom section is for **Codec priorities**. We will discuss codecs in Chapter 7, *Enterprise Features*.

We also see a **Which IP to use in 'Contact' field for registration** setting in the **Registration Settings** section. If you have a static public IP, you can set it here. If you are using a dynamic IP, you will want to leave the default setting as **External (STUN resolved)**. STUN is **Simple Traversal of UDP through NAT**. We will cover NAT in Chapter 10. Some ITSPs support STUN while others do not. For a business, you will probably have a static IP. Most home Internet connections will be dynamic, and you will get better results if you have STUN.

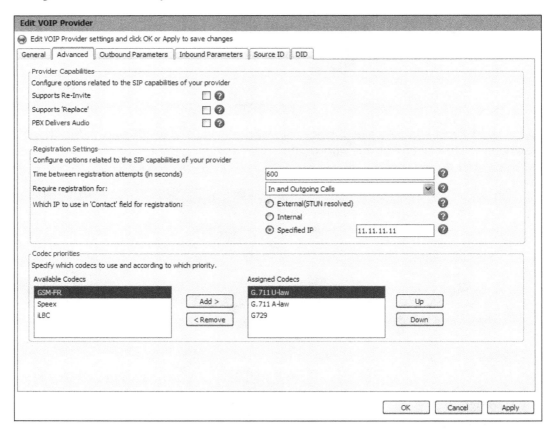

Now go ahead and click the line information for this VoIP provider in the left-hand window pane. The VoIP provider name is for all the lines under this account. If we had a carrier that did not support concurrent calls, we would have had to sign up for more lines with the same carrier. We can then specify what to do with each line:

In the right-hand pane, we now see the details for this particular phone number. The only thing I want to change from the default is the **Outbound Caller ID**. As this provider supports what you want to use for caller ID information, I can specify my name here. Click **Apply** when you are done with these changes:

Edit VOIP Provider

▶ **Edit Port :123456789**

Voip Provider port no 123456789 on Callcentric

Virtual Extension Number (Unique)	10008	⑦
Authentication ID	mycallcentricusername	⑦
Authentication Password	••••••••••	⑦
Port identification	123456789	⑦

Office Hours
Configure where calls should be routed during office hours.

○ End Call		
○ Connect to Extension	100 Rob Lloyd ⌄	⑦
⦿ Connect to Queue / Ring Group	802 Sales ⌄	⑦
○ Connect to Digital Receptionist	801 DR 1 ⌄	⑦
○ Voicemail box for Extension	100 Rob Lloyd ⌄	⑦
○ Forward to Outside Number		⑦
○ Send fax to email of extension	888 Default FAX Destination ⌄	⑦

☐ same as during office hours

Outside Office Hours
Configure where calls should be routed outside office hours.

○ End Call		
○ Connect to Extension	100 Rob Lloyd ⌄	⑦
○ Connect to Queue / Ring Group	800 MainRingGroup ⌄	⑦
○ Connect to Digital Receptionist	801 DR 1 ⌄	⑦
⦿ Voicemail box for Extension	104 Zachary Alan ⌄	⑦
○ Forward to Outside Number		⑦
○ Send fax to email of extension	888 Default FAX Destination ⌄	⑦

Other Options
Set other general options for this specific port

Outbound Caller ID	RobLloyd	⑦
Allow outbound calls on this line	☑ ⑦	
Allow incoming calls on this line	☑ ⑦	
Maximum simultaneous calls	3	⑦

[OK] [Cancel] [Apply]

We have finished the SIP trunk settings. Now, we need to assign a **DID number** to the SIP trunk. This number is used to identify the line and helps us create inbound rules. On the VoIP providers settings, there is a tab named **DID**. Add your assigned DID number(s) to this trunk by entering them, and then click **Add**. When you are done entering the DID number(s), click **OK** or **Apply** at the bottom to save them. Now, give it a try and see if you can make outbound calls.

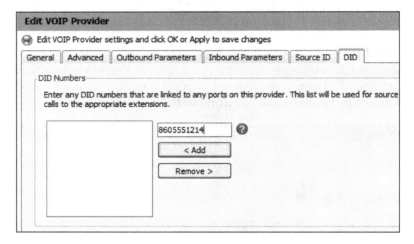

Summary

In this chapter, we discussed why you need a trunk, what you need for a PSTN line, what to look for in a VoIP provider, the equipment needed, analog lines, call quality, disaster recovery, and finally how to integrate a trunk into 3CX using the easy-to-use wizards.

That is a lot of information, but it's all needed to connect to anyone outside your network.

6
Configuration

This chapter is all about the configuration options available in 3CX. It's these little configurations that give us a better, more professional, and easier-to-use phone system. In Chapter 4, *Call Control: Ring Groups, Auto-attendants, and Call Queues*, we discussed using call queues and using special Music on Hold (MOH) for creating custom waiting messages or announcements. We are going to discuss these features in detail and show you even more options that we can use. We are going to cover these topics:

- Hold music
- Prompt sets
- Dial plans
- Direct extension calling

Music on Hold

Music on Hold (MOH) is already included in the phone system installation. It's a royalty free music file that sounds fine for most people. To create a custom sound, we will need to change the file. How do we get this new file? Read on!

At the moment, the 3CX Phone System's Music on Hold will simply play one music file. There are no play lists, volume setting, auto-mixing, or other advanced Music on Hold features. Although there have been a lot of community requests for this and many people wonder if they can plug their MP3 player into the audio IN jack on their PC for Music on Hold, currently it is not in the plans and 3CX Phone System does not support it.

Obtaining the file

If you have a favorite song that you would like to use, you will need to get permission to use it for commercial purposes. After all, it is copyrighted and you wouldn't want to steal it, right? You need to contact the artist and get permission for it. If it's for home use or a very small business, you might be able to get away with using it and never having a legal issue. However, if you are a large company, maybe even with a global presence, you should always get permission.

However, what do you do if you cannot get permission to use the music you want? You can either create it yourself or look for royalty free music. Do you have a corporate jingle that you used in a TV commercial? Use the same jingle and get that corporate branding working for you. If you don't have a jingle, you can hire that local garage band, or bribe them with some pizza to create a corporate sound for you.

No corporate jingle? No local band? Google is your friend; do a search for royalty free music. Keep in mind that these sites do not provide "free" music. They are "royalty free", which means that you do not have to pay for each use. Some of the music is affordable, and you can use it forever. Try these two sites: http://www.royaltyfreemusic.com and http://www.neosounds.com.

For the iTunes user

For now we can assume that you have permission to use a track off of a CD. We need to get that track off of the CD and into the correct format for 3CX. To do this, you will need some audio ripping software. As iTunes is extremely popular, I will show you how to record the track using the correct settings:

1. Start by opening iTunes. I'm using the latest version at the time of writing this book—8.2.

2. Click **Edit | Preferences**. On the **General** tab, under the section that says **When you insert a CD**, select **Ask to Import CD** and click the **Import Settings** button as shown in the following screenshot:

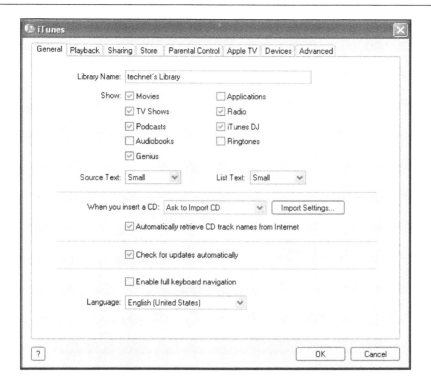

3. On the **Import Settings** screen that appears, select **MP3 Encoder** from the **Import Using** drop-down list. Under **Setting**, select **Custom** from the drop-down list:

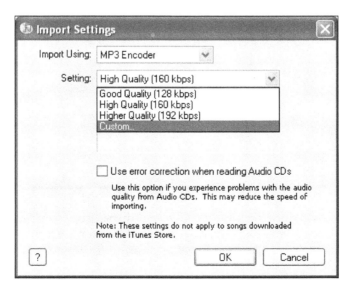

4. Now an **MP3 Encoder** window will pop up. Adjust the following settings:
    ◦ **Stereo Bit Rate: 16 kbps**
    ◦ **Sample Rate: 8.000 kHz**
    ◦ **Channels: Mono**

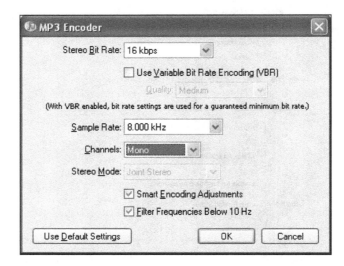

5. Click **OK** and save the settings. You will need to change these settings back to default or as per your preferences for normal CD ripping. While these settings will be perfect for a phone system, they won't sound good on your iPod.

6. Insert the CD you want to rip a track from, and iTunes will prompt you to import the CD, as shown in the following screenshot. Click **Yes** and ripping will start.

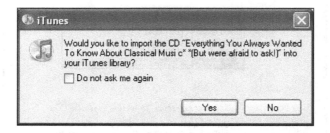

7. After iTunes is done ripping the CD, edit the settings for the call queue that we made back in the *Call queues* section of Chapter 4. At the bottom, you will see **Other Options**. Click the **Browse...** button for the **Music on hold** setting:

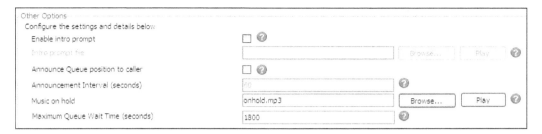

8. The **Choose a file -- Webpage Dialog** window will open. Click **Add**:

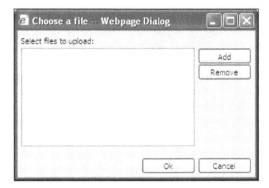

9. Next, you'll have to navigate to your music file. The default folders for iTunes are My Documents, My Music, iTunes, iTunes Music. In one of these folders you will see the .mp3 files you created. Select the one you want and click **Open**:

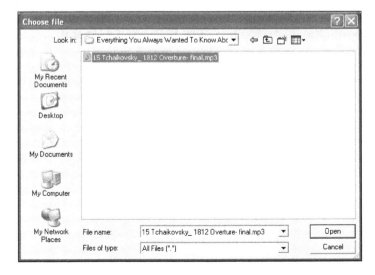

10. Once you see your file in the **Choose a file** dialog box, click **OK**:

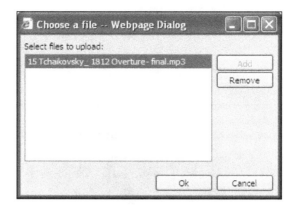

11. You will now see your updated MOH file in place of the default. Click **Apply** to save the settings:

12. Now you are done. You can use the same file for your standard MOH file as well, if you like. To change those settings you will need to adjust the **General Settings**. On the menu toolbar, click **Settings | General**:

13. You will now see an MOH section. Click **Browse** and select the file you would like for your standard MOH file. Click **OK** when you're done:

That's it for the Music on Hold settings. Just remember that it's best to use something soothing. If you're into customer service for disgruntled customers, you don't want music that will pump them up even more! I have a client who does estate work for elderly people; they thought the default music that 3CX comes with was too crazy! I changed it to some classical music, and the seniors didn't mind being on hold anymore.

Prompt sets

The default language for 3CX is English. Unless you change it, everything that is announced is in English, such as *Please hold while I transfer your call*. If you want it in another language, you will need to download the language files as follows:

1. Click **Links | Phone System updates**:

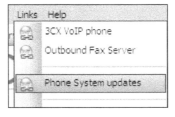

2. On the bottom of the screen, you will see the section **System Prompts Update**. Place a checkmark in the checkbox next to the language that you'd like to use, and click **Download Selected**. I'm going to pick Italian as an example:

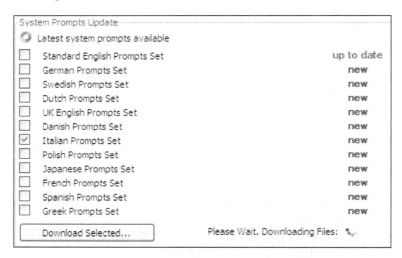

3. To use the new language, you will need to go to **Settings** and click **System Prompts**:

4. On the **System Prompts Settings** screen, click **Manage Promptsets** and another dialog box will open:

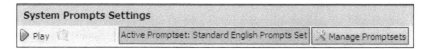

5. You will see a couple of options here. Click on the drop-down list next to **Active Promptset** and choose the new language as shown in the following screenshot.

6. Then, click the **Set As Current Promptset** button:

7. Once you click **OK**, your descriptions and filenames will change to the new language that you selected:

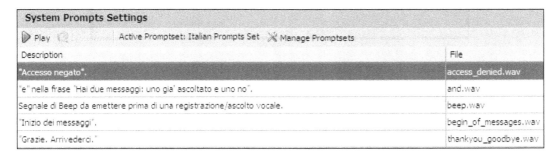

8. If you want to have your own personalized prompts, you'll need to copy a prompt set. Click the **Manage Promptset** button again and click **Copy Promptset**:

9. Give it a name that you want to use and click **OK**, as shown in the following screenshot. This will copy the language prompts that you have highlighted and save them as new files. You can change the default files, so this is necessary:

10. Now you can change your **Active Promptset** to the one you copied:

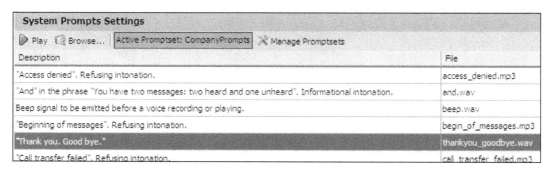

Now you will see the **Browse** button become active that was grayed out previously. With the new prompt set, you can upload your own recordings. Click the description/file you would like to change and click **Browse**. Then, go find your custom recording file. These files are recorded just like the one for the Digital Receptionist. You can refer to Chapter 4, the *Digital Receptionist setup* section, if you don't remember how to do that.

 Although 3CX Phone System supports multiple languages, it only supports one at a time. For example, some companies have a prompt to "Press 1 for English and 2 for Spanish" and from there on, all the phone system prompts change to a different language. 3CX does not support this scenario.

Dial plans

In Chapters 4 and 5, we worked with dial plans. Let's tear into them and create some custom plans that we can use with some tips for limiting access.

The next screenshot is of the basic dial plan that we created for the analog trunk line. As a quick review let's go over it again.

In the first section **General**, we see a **Rule Name** field to enter the name for the rule. Then we see the section **Apply this rule to these calls**, which has the following options:

- **Calls to numbers starting with (Prefix)**: 9
- **Calls from extension(s)**: 100-102
- **Calls to Numbers with a length of**: 3,7,10,11

In the section **Make outbound calls on,** we see which outbound line to use but for now, we are just using the analog gateway.

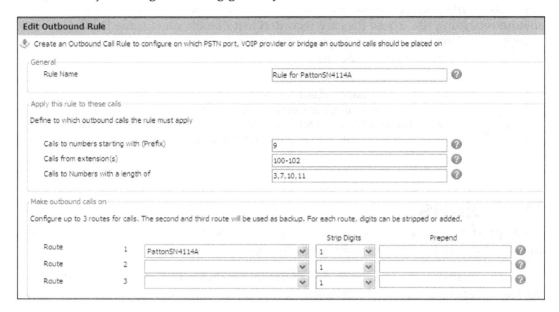

Let's break this down in more detail.

The **Calls to numbers starting with (Prefix)** option will help you separate what calls go out on what lines if you have several outgoing lines. If you have unlimited local calls using the PSTN line, you may want this to be the default one that everyone uses — tell them to dial "9."

What if they dial "9" and then start a long distance call? This line won't accept it! It will only dial numbers up to eleven digits. For example, 9,123-555-1212 would be a local call if they had to use the area code. Now you have just saved yourself some money on a long distance call that you can do less expensively using our VoIP line.

Using our Callcentric VoIP line outbound rule, we can see that any call with twelve digits will go out on this line—8, 1-456-555-1212. Now we just need to strip off the "8" before it goes out of the system:

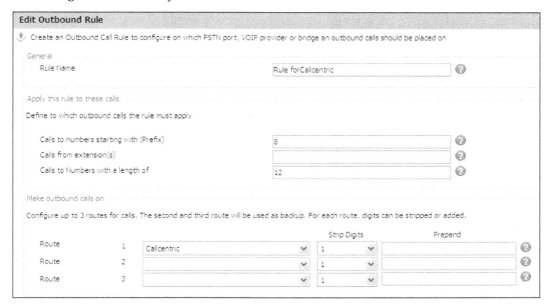

What about disaster recovery? What if the Callcentric line is down? Well, now we might want to use the more expensive analog line. Add a new **Route** to the call so that if the first one fails, it will use the second route:

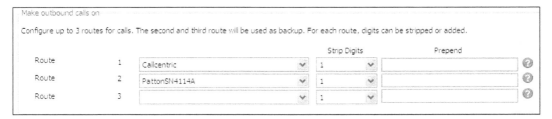

What about controlling access to lines? With a little number massaging, we can create "passwords" to use for an outbound line.

 Please note that "line passwords" are not a feature of 3CX and are not supported.

We will still use the Callcentric line, but we can set it for controlling international calls. Let's say we want to call St. Andrews and make a tee time. We need to call 011 44 01334 466666. Our user will dial a password that we have set as 468 for "INT." So, this person will dial 468-011 44 01334 466666. Just remember to change the "password" once in a while, in case it leaks out.

The following screenshot shows what the rule will look like:

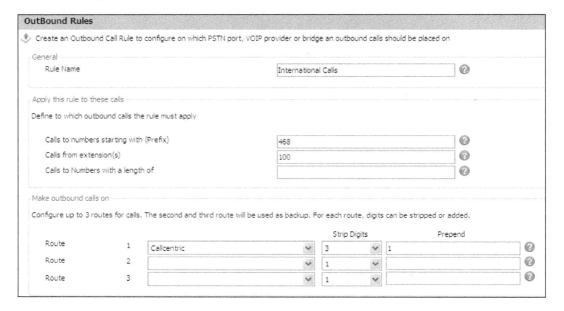

How does it work? Well 3CX sees the 468, so it looks at this call and routes it to this outbound rule. We only want to have extension 100 to be able to use this for even more security. 3CX now has to strip off the 468, so we strip three digits. Callcentric still needs all calls to start with a "1" so we can **Prepend** the "1" and not have the user dial that as well.

With some creativity, you can create some very sophisticated outbound rules. Just make sure that whatever you create still lets anyone dial emergency services.

Direct Inward Dialing (DID)

It's time for some inbound rules. We currently have four lines on the analog line that we can direct to specific extensions.

The first one we will create is for any call coming in on line 4 to my extension. Click the **Create DID** button on the top toolbar:

First, we give it a name. Then in the **DID/DDI number/mask** field, we can put a "*." This tells 3CX to use this for any call. Now we specify line **10003** in the **for port identification** field.

Next we tell it where to direct the call which, in this case, is extension 100. We want this to happen outside of our office hours, so we can check that box as well, as shown in the following screenshot:

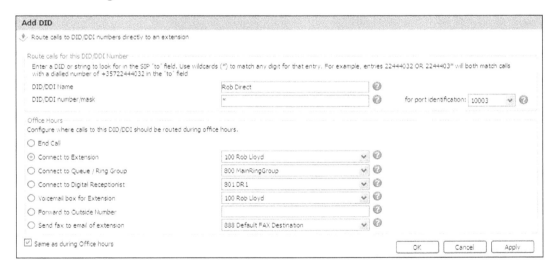

Simple enough, right? Let's create a better one—this rule will inform me that my wife is calling, avoid the receptionist, and go right to my office.

Create a new DID. Now instead of the "*" in the **DID mask** field, we enter the phone number that we want to use as a filter and apply this to the main line. If it's after hours, I'd like this to be forwarded to my cell phone. Don't forget your outbound rules—I added a "9" to use the analog outbound lines.

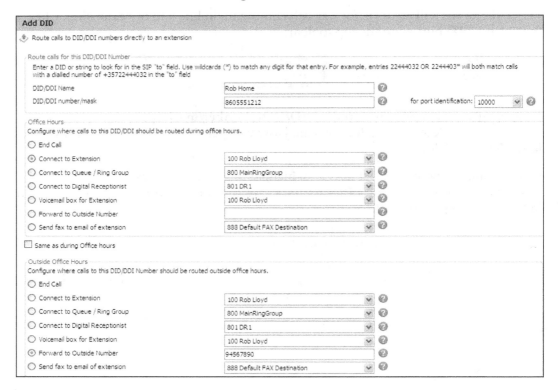

Summary

We covered a few advanced features in this chapter. We created custom Music On Hold, specific outbound rules, prompt sets, and direct dials, all of which give you more flexibility in a professional sounding, flexible, and powerful phone system.

7
Enterprise Features

In this chapter, we will cover some of the great enterprise features available in 3CX and some semi-advanced networking topics dealing with remote phones. These features are not available in the free version, so you will have to pay a little for the enterprise license. It's worth it!

This chapter's topics include the following:

- Remote phones
- Remote phones using a VPN
- Networking topics, such as NAT, port forwarding, and configuring a Linksys router
- Call recording
- Conferencing
- Call reporting
- Faxing
- Codecs

Remote phones

One of the great features of an IP PBX is the ability to easily connect remote phones to your 3CX IP PBX using existing networking infrastructure. In fact, I find that a very common reason for organizations replacing their existing phone infrastructure is being able to connect remote phones or branch offices with several phones.

There are several ways to connect remote phones to your 3CX IP PBX, such as establishing a hardware **Virtual Private Network (VPN)** between your remote phone and 3CX server site, establishing a VPN between a VPN-capable phone and your 3CX server site, and port forwarding your VoIP traffic through your firewall.

I have been able to achieve the most consistent and headache-free results by using the VPN methods to connect remote phones. Using a VPN avoids many issues that can result from attempting to traverse NATs and firewalls that are hard to get, to handle SIP packets correctly, broadband modems that double NAT networks and ISP's that block VoIP traffic.

Remote site to 3CX site VPN tunnel

A site-to-site VPN will allow you to set up phones just like they are on your main 3CX server site. We won't go into the details of setting up a site-to-site VPN, as that is a standard networking skill you can find elsewhere, but you can see the overall setup in the following figure. I will note that using a VPN will have the added benefit of securing your VoIP conversations which is becoming increasingly important. A VPN will also require slightly more bandwidth overhead. With the site-to-site VPN method of connecting remote phones, it is much simpler to add more phones and avoid NAT issues entirely.

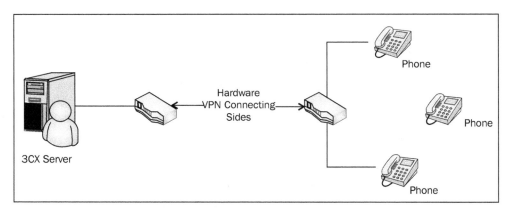

VPN-capable SIP phone to 3CX site VPN tunnel

Snom made the Snom 370, which includes a built-in OpenVPN client. As of this writing, this is the only VPN-capable 3CX supported phone that I know of. The remote phone connection method is beneficial because you don't need to "unsettle" the existing firewall infrastructure that may be in place at your organization, as the remote phone has a dependable and secure connection to the 3CX server.

The VPN-capable SIP phone method of remote connection is shown in the following figure:

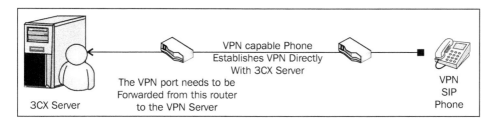

 The Snom 370 includes an OpenVPN client that can only connect to an OpenVPN server. OpenVPN server software is available for free. An OpenVPN client cannot connect to a **Point-to-Point Tunneling Protocol** (Windows Server VPN implementation) or **Internet Protocol Security** (many mainstream hardware firewalls) VPN server.

The Snom 370 is the only business class phone I am aware of that includes a VPN client built into the phone. This makes an excellent secure, remote phone.

The Snom 370 phone includes a built-in OpenVPN client that can be used to establish a VPN to any OpenVPN server. OpenVPN is free and can be installed on a Windows computer, including the 3CX server itself. Mike Harris at Worksighted Inc, has written an excellent step-by-step document to guide you through this, admittedly, complex initial setup process. The second remote has far fewer steps, so don't be daunted. The following blog post will have a link to the tutorial:

`http://3cxblog.worksighted.com/2008/09/first-post-test.html.`

I would like to thank Mike Harris for these clear and detailed instructions.

Several benefits of using this method include the security of your VoIP traffic, the absence of NAT traversal issues, and the mobility of the remote phone, which can be taken to any location with Internet access.

While the initial configuration has quite a few steps, once you have the infrastructure in place, the next remote phone is much less work.

Port forwarding method to connect a remote phone

Port forwarding is another method to connect a remote phone to your 3CX phone server. If all the pieces fall in place for you, this method can be incredibly simple. If you hit a snag, it can be one of the most time-consuming parts of your IP PBX implementation!

> Port forwarding can be the most trouble-prone way to connect a remote phone to the 3CX server. It is subject to problems issuing from NAT traversal, ISPs blocking VoIP traffic on certain ports, and difficulty in correctly configuring routers to forward VoIP ports. If you get stuck trying to get this method to work, I suggest trying one of the VPN methods.

With this method, our VoIP traffic is directed from our remote phone over the Internet, through our firewall, and finally to our 3CX server. Each leg of this path is a minefield of possible trouble spots; so before we start, let's double-check the following:

- Make sure that the remote site firewall is not blocking any VoIP ports. Typically, outgoing ports are not blocked, but verify that nothing is being blocked on the remote site.

- Make sure that our ISPs are not blocking any VoIP ports. If any of the ISPs involved in carrying our traffic block VoIP ports, then our remote phone will fail. The only way to check this is to call the ISP.

- Make sure that the 3CX server is not behind double NAT. One of the common reasons a network is double NATed is that the modem is also a firewall, and then there is a dedicated firewall that is also doing NAT. Make sure your traffic only goes through one NAT. Believe me, it's hard enough to get things working with one NAT!

- Make sure that your firewall does static port mappings (otherwise known as full-coned NAT). A very firewall that does static port mapping is the Linksys WRT54G, as shown in the following image. It is a simple, low cost unit, and we will use it in our example setup of 3CX and a remote phone.

The Linksys WRT54G router does static port mapping that is required to use the port forwarding method of the remote phone setup. Make sure your firewall supports static port mapping.

For more information on NAT, check out `http://wiki.3cx.com/documentation/networking/nat-firewalls`.

Following is a diagram of a port forwarding remote phone scenario:

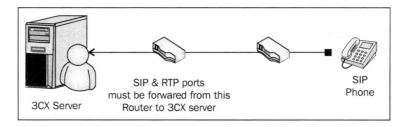

 The following link will help you configure your Linksys router for the 3CX Phone System: http://wiki.3cx.com/documentation/networking/linksys-configuration

To traverse the firewall, we will need to enter the port forwarding information in our Linksys WRT54G router. First, we will need to log in to the web interface of the router. We will click on the **Applications & Gaming** tab and then the **Port Range Forward** tab, as shown in the following screenshot:

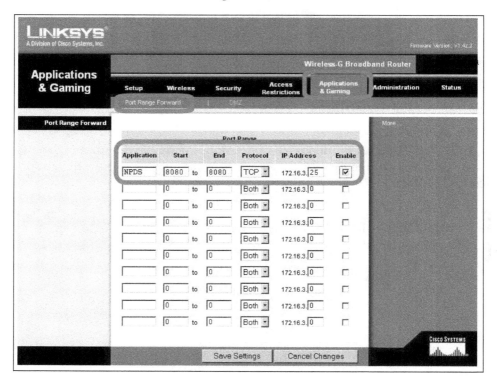

The IP address column should be the IP address of the 3CX IP PBX server. Following are the ports that need to be forwarded:

- **Port 5060 (TCP and UDP)**: Used for SIP protocol transmission
- Port 5090 (TCP): Used for 3CX tunnel (if tunnel is enabled)
- **Port 9000-9015 (UDP)**: Used for RTP for incoming and outgoing calls

After we click **Save Settings**, our WRT54G setup should be complete and our remote phone should work.

Using the 3CX Firewall Checker

If you are having problems with the port forwarding method, 3CX provides a diagnostic tool to detect what is wrong with port forwarding. To use the 3CX Firewall Checker, log in to 3CX. Using the **Services Status** screen in **3CX Management Console,** stop the **3CX Phone System** and **3CX Phone System SIP/RTP Tunneling Proxy** services. Then, click on **Settings | Firewall Checker**:

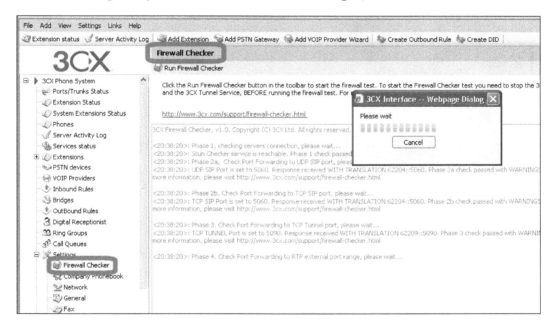

Click on **Run Firewall Checker** to start the test. Each port will be displayed with a code to tell you its status. You can check on 3CX's web site at `http://www.3cx.com/support/firewallchecker.html` to see what the error codes mean.

Port forwarding using the 3CX SIP proxy tunnel manager

This is a slight variation of the port forwarding method, which adds a tunnel manager utility into the mix. The SIP proxy tunnel manager is a software that runs on any Windows PC that will cause all the SIP and RTP traffic to tunnel through one port. This solves several problems with the port forward method:

- Get around ISPs that block certain VoIP ports
- Simplify router configuration as there is only one port to forward
- Resolve the NAT issues
- Work around hard-to-configure routers such as Sonicwall

[You will find a detailed documentation about the SIP proxy manager at `http://wiki.3cx.com/documentation/networking/sip-proxy-manager`]

The following figure illustrates how to use the tunneling method to connect a remote phone:

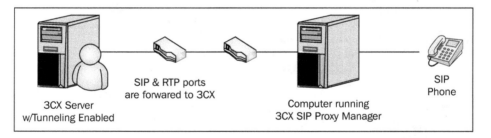

The first step in setting up a proxy tunnel is to enable the **3CX Tunnel** by logging into the **3CX Management Console**, clicking on **Settings | Network,** and then the **3CX Tunnel** tab, as shown in the following screenshot. Enter the **Tunnel password** and the **Local IP** that we want to use, let the port default be **5090**, and then click **Apply**. Now, we are done configuring the 3CX tunnel:

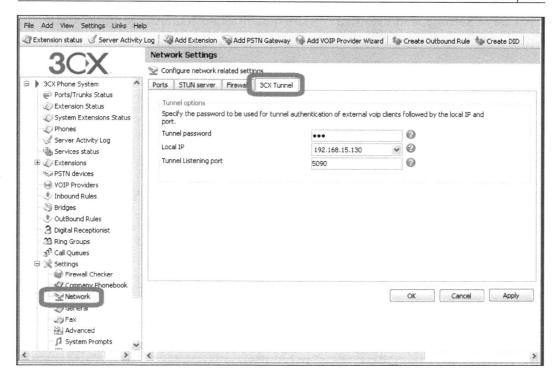

Next, we need to set up port forwarding in our router as we noted in the last section. We only need to forward Port 5090 (TCP).

Now we need to install the 3CX SIP Proxy Manager at the remote site. The install file can be found at `http://wiki.3cx.com/documentation/networking/sip-proxy-manager`.

After we have installed the proxy, we can run it to configure it using the next screenshot as a reference.

- **SIP Listener IP Address** is the IP address of the remote PC the proxy is running on
- **SIP Listener Port** can be left as 5080
- **Server Public IP Address** is the public IP address of the 3CX server
- **Server Tunnel Port** is the port that we set up in the 3CX server
- **Server Tunnel Password** is the password that we set up in the 3CX server

Click **Save Settings** and then **Exit**. We are now finished configuring the **3CX SIP Proxy Manager**, and we will move on to configuring the remote phone.

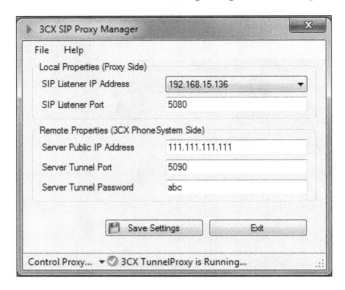

The last thing we need to do is configure a phone to connect to 3CX using the 3CX proxy tunnel. For our example, we will use the 3CX softphone, as shown in the following screenshot. Please note that some hardware phones also support using the 3CX SIP proxy tunnel, such as the Grandstream GPX series phone.

 Some SIP phones are known NOT to work with the 3CX proxy tunnel. Aastra phones do not work with the tunnel protocol.

Open the 3CX softphone and open **Connection settings**. Check **I am out of the office** and set the **extern IP** to the public IP address of our 3CX server. Check **Use tunnel**, set **Local IP of remote PBX** to be the IP address of the 3CX SIP Proxy Manager (**SIP Listener IP Address**), and set the **Tunnel password** to be the password we set up in the 3CX server. All other phone configurations can be set just like it is local to the 3CX server. Click **OK** and we should be done:

Call recording

Call recording is a great way to cover your backside. If you promised one thing and the customer expects something else, here is your proof. Recordings are also useful for training new customer service representatives or sales people. Record their calls and evaluate and educate them on how to do better. Or, if the call was perfect, use the call for training others.

There are some legal issues with recording calls. Some laws prohibit recording calls without the other person knowing it. Some laws require a "beep" tone every minute so that you know the call is still being recorded. Check your local laws before recording any call.

Call recording can be done in a couple of ways once you purchase a license. On the 3CX VoIP Phone software, there is a Record button. There is also a Record button on Snom phones. Both do the same thing, which is starting and stopping the recording of the calls.

Conferencing

If you want to conference your calls, you must buy a license. This is not available in the free edition. While there are many paid-for conference call options, VoIP providers offer them Skype, GoToMeeting, and many others. However, it's easier and cheaper to do your own through 3CX.

Conference calls can be set up as you need them. No need to define them and have them sit around doing nothing.

You are limited to 32 callers in total, using up to 8 conference calls on your system. The license you buy will determine how many you can have. Buy your license with this in mind if you have the need for a lot of conference calls.

Conferencing is enabled by default with an extension number of 700, allowing 4 conferences to be going on at the same time. You can change these settings from the **Settings | Advanced Settings | Conferencing** tab.

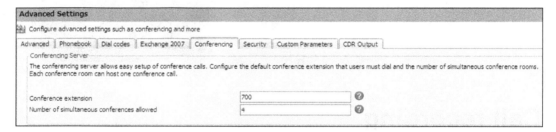

The more simultaneous conferences you have, the more processing power you will need. This is not something you'd want to do on a virtual PBX system. I would dedicate a full system with the latest processor to handle a large conference system.

Creating a conference call

To create a conference call, you will need to dial 700 to get started. It will ask you for a conference ID. You can use any number you'd like, but everyone who wants to join the conference call will need to use the same number.

If you are the first person creating the call, you will be asked to confirm the creation of the call. Press * to confirm or # to cancel.

Once that is done, it will ask you to record your name. Once you are in the conference call, you will get hold music while you wait for others to join. When they join, their names will be announced.

That's it! It's a very simple process to set up ad hoc conference meetings.

Call reporting

In previous versions, there were built-in reports from the main 3CX interface. Now that reporting has gotten bigger and better, was moved to a separate program. **Under Programs | 3CX Phone System**, you will see a **3CX Log Reporter** utility.

Run this program, and you will have a menu of choices based on what type of reports you'd like to have:

- Call logs
- Call statistics
- Queue statistics
- Dropped/Abandoned calls
- Agent statistics
- Ring group statistics

Run the reports that you like. Some of them should be viewed daily, some weekly, and some monthly. Statistics report is one of the daily reports to check out. This will help you determine if you have any issues to tackle. Others, like the abandoned calls, might also be helpful on a daily or weekly basis. Queues, Agents, and Ring groups will only be helpful if you use those features. I'd suggest looking at each one to see what is offered.

Faxing with 3CX

The fax feature is another paid feature that gives you a full fax-to-e-mail server using the standard T.38 fax format. The gateway line must also support T.38. Most VoIP providers do not support faxing, so check with them before signing up if this is a needed option.

When a fax comes into 3CX, it goes to the fax server portion of the system. It will convert the fax into a PDF and then e-mail it to the fax recipient.

You will need a dedicated line to use as a fax line, or you can set up a DID as a fax line over the same VoIP line (using a dedicated number for a fax).

To set this up you will need to specify a DID line, like we did earlier in this chapter. Only now, we need to specify the rule to the **Send fax to email of extension** field. You can use the **888 Default FAX Destination**, or you can use the e-mail of any extension you have created for dedicated fax lines:

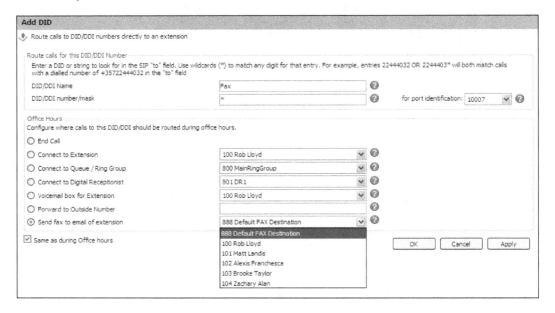

If you would like to change the fax server e-mail address, go to **Settings | Fax**. This will bring you to the screen where you can change the fax server extension number, ID, and password for the fax server to log in to the SIP server like any other extension does. You can also define or change the e-mail address of the fax recipient here:

Once you change these settings, you will need to restart the fax server service for them to take effect.

Codecs

Codecs are used to convert analog signals (human voice) to digital signals that can be transmitted over some network medium, such as CAT5, fiber, wireless, and so on.

Each codec uses a certain amount of network bandwidth. The amount of bandwidth you have available will determine how many calls you can have at one time before your calls start having problems. Let's say you have the standard $30/month DSL at your house. This can typically be 5MBps of download speed but only 700Kbps of upload speed. Sure, you can download files quickly, but you can't send them. This is an important concept with VoIP. Each codec will use a certain amount of bandwidth to pass the digital data along. The standard is ITU G.711, also called A-law or U-law. This has only a little bit of compression on the line and is used most often because of the call quality. Another popular codec is ITU G.729. This codec is one of the better ones with regard to call quality and bandwidth use. The biggest problem is that you have to pay for it. 3CX (and any other IP PBX system) has to buy the license from the owner. They, of course, pass this extra cost on to us.

Here is a list of the available codecs and the amount of bandwidth they use:

Codec	Bandwidth
Speex	2.15 to 44.2 Kbps
LPC10	2.5 Kbps
DoD CELP	4.8 Kbps
ITU G.723.1	5.3/6.3 Kbps, 30ms frame size
ITU G.729	8 Kbps, 10ms frame size
GSM	13 Kbps (full rate), 20ms frame size
iLBC	15 Kbps 20ms frame size, 13.3 Kbps 30ms frame size
ITU G.726	16/24/32/40 Kbps
ITU G.728	16 Kbps
ITU G.722	48/56/64 Kbps
ITU G.711 (also known as A-law or U-law)	64 Kbps

As you can see, there are plenty to choose from, and planning is critical to determine the ones you can use and how many simultaneous calls you can support. Our 700Kbps DSL line will support about seven simultaneous calls using the default G.711 codec. If we can use GSM now, we can support about 35 simultaneous calls. That is correct; I didn't do an exact math calculation here. You certainly don't want to plan on maximizing your bandwidth. You will need some extra room for other traffic and overhead to support the line.

This rough estimate will help guide you in figuring out what you need for an Internet connection. If you have 200 users and predict 100 of them will be on the phone at the same time, then a 10/100 Ethernet line will be full once you add in e-mail, files, and web traffic. Now, you have a couple of choices: subnet your networks or go all gigabit from the servers, switches, phones, desktops, and so on. What about wireless? They have wireless phones, but will your office support all the wireless devices and VoIP at the same time?

So, what codec(s) can you use? That depends on your processing power and what your VoIP carrier offers. Almost all of them support G.711, while some support G.729 and GSM.

If you aren't familiar with compression, think of it as taking data and dropping enough here and there so that the picture is still the same from a distance but, up close, some pixels are missing or the edges are jagged. Voice has the same effect—drop a little off the higher end, drop a little off the lower end, drop very light signals, and process it all on both ends. Some of them have silence compression. In other words, if you aren't speaking, it will not transmit any voice packets. When you start to talk, it might drop a couple packets of information while it turns back on.

If you can fully support the default G.711, you should be fine and all your calls will sound great, possibly even better than a traditional PSTN telephone.

Remember that compressing the calls takes some power, so don't just go with a wimpy processor and expect perfect calls. It will work for a couple of calls but not for a 50-user office. When you buy 3CX, you get a few licenses of G.729. The larger the license, the more you get, but only up to 16 simultaneous calls.

Once you find out what codec your provider supports and if you have paid for a 3CX license, then you can use G.729 or any of the others.

Edit the settings of your VoIP provider. On the **Advanced** tab, you will see **Codec priorities**. Here, you can **Add** or **Remove** codecs and change the default order:

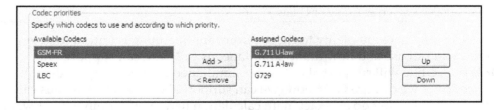

Summary

This chapter covered some of the advanced, mostly paid-for features of 3CX. Getting the G.729 codec and faxing are the top two reasons to have a licensed version of 3CX. Surely, queues, call recordings, and everything else is great, but if you are limited in bandwidth, this can be the difference between a working system and one that no one wants to use because the call quality suffers.

8
3CX Integration

In the last chapter, we looked at some very interesting characteristics of 3CX. As 3CX is based on standards and non-proprietary protocols, it can be integrated with many other standard-based software and devices, as well as analog devices. In this chapter, we won't discuss every integration possibility with 3CX, but we will go through some common ones to get your mind going. We will first explore connecting 3CX to Microsoft Outlook with the free TAPI dialer. Next, we will show you how to tightly integrate 3CX with a free instant messaging server. We will wrap up by showing you how to link a Legacy PBX with 3CX so that you can migrate to new technology at your own pace.

We will take a look at the following:

- Outlook Click-to-Dial integration
- Instant Message Server integration
- Installing and integrating Openfire Instant Message Server
- Integrating a Legacy phone system
- Some more integration possibilities

Outlook 2007 Click-to-Dial integration

3CX includes a TAPI driver that will allow Microsoft Outlook 2007 to dial phone numbers directly from inside Outlook. The TAPI driver is installed when we install the 3CX Call Assistant software. After you have installed and configured the 3CX Call Assistant, launch Outlook. Open a **Contact** as shown in the following screenshot, make sure there is a valid phone number entered and click on the **Call** button:

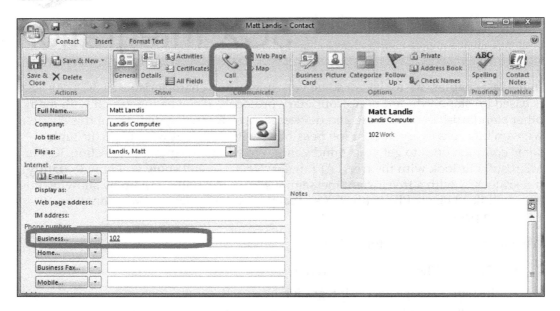

You will be presented with the **New Call** window. Here we will click **Dialing Options** to change the TAPI driver to the newly installed 3CX TAPI driver:

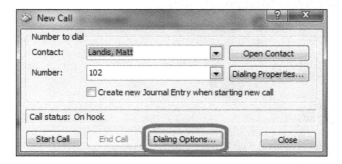

You will now see the **Dialing Options** window. Set the **Connect using line** drop-down to the **3CX Call Assistant** option and then click **OK**:

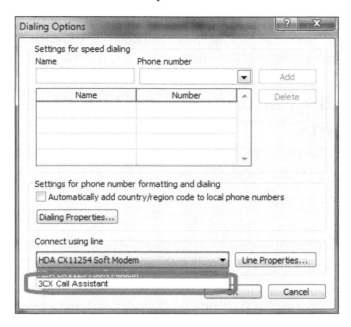

Outlook is now set up to be able to dial a phone number using 3CX.

Integrating Instant Messaging Server

One of the increasingly demanded pieces of a **unified communication system** is instant messaging functionality. In the Windows world, perhaps the most well known corporate instant messaging server is **Microsoft Office Communications Server** (**OCS**). OCS started with instant messaging as its foundation and is constantly adding PBX features, while other traditional PBX products start with phone features and then add instant messaging as a feature.

As 3CX is a Windows product and Microsoft makes an instant message server, why not walk-through integrating 3CX and Microsoft OCS? As noted earlier in this book, 3CX has made a decision not to support this integration because OCS appears to be heading in the direction of direct competition with products such as 3CX. While Openfire is not officially supported either, it is not on 3CX's "we will not support" list and is considered a workaround until instant message capabilities find their way into 3CX.

At this point, 3CX includes a simple instant messaging server but does not include features such as group chat, chat archiving, or federation with public IM services such as MSN or Google Talk. As 3CX is based on the standard SIP protocol, it can be integrated very tightly with other full featured instant messaging systems amazingly easily. One of those is the open source product **Openfire**. We will walk-through the process of integrating 3CX with Openfire.

Downloading and installing Openfire components

Openfire is a server component, and Spark is the client. You can download these components at `http://www.igniterealtime.org/downloads/index.jsp`.

Install the Openfire server using the documentation located at:

`http://www.igniterealtime.org/projects/openfire/documentation.jsp`.

Openfire is quick and very to simple install and can be done in 15 minutes, if all goes well. You can install the Openfire server software directly on your 3CX server.

After Openfire is successfully installed, the startup screen will open as shown in the following screenshot. You can then click on the **Launch Admin** button to go through the initial Setup wizard:

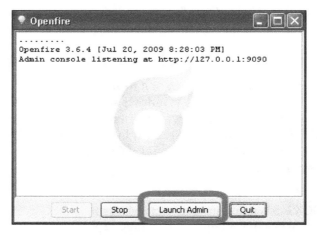

A browser will open and run you through the setup options. When you get to the **Database Settings** screen, select **Embedded Database**:

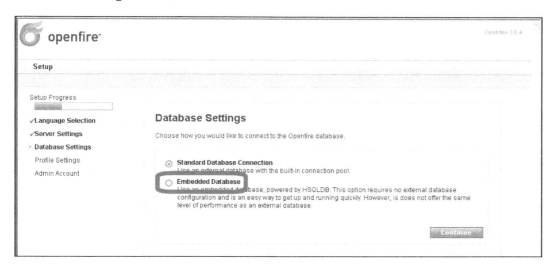

On the **Profile Settings** screen, select **Default**, as shown in the following screenshot:

 If you have a Windows Server, you can integrate with Active Directory, as 3CX does not integrate with Active Directory at this point.

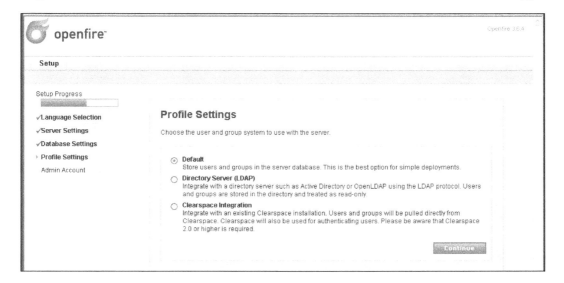

When you are done with the initial settings, you can click **Login to the admin console** to configure Openfire:

You can now log in to the web **Administration Console** using the password you assigned to the Openfire administrator:

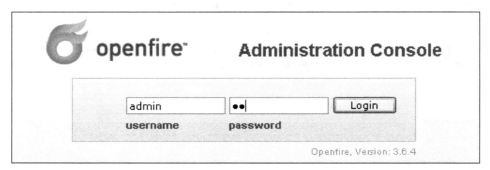

Basic Openfire configuration and Spark install

Now, we are ready to do the basic configuration of Openfire. We'll configure some users in Openfire by clicking **Users/Groups | Create New User**. Let's add several users that mirror the users in 3CX. Once you have added several users in Openfire, it is ready to be used as an instant message system. Now, we will integrate it with 3CX:

Integrating 3CX and Openfire

The first step in tying 3CX and Openfire together will be adding some plugin to the Openfire server, so let's log in to the **Administrator Console**.

 For the plugin installation, the Openfire server needs to have Internet access.

Now click on **Plugins | Available Plugins**. Then install the **Client Control** and **SIP Phone Plugin**, as shown in the following screenshot:

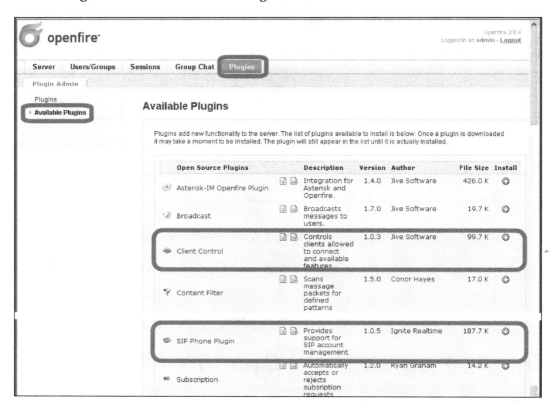

Let's verify that the plugin is installed by clicking on **Plugins**:

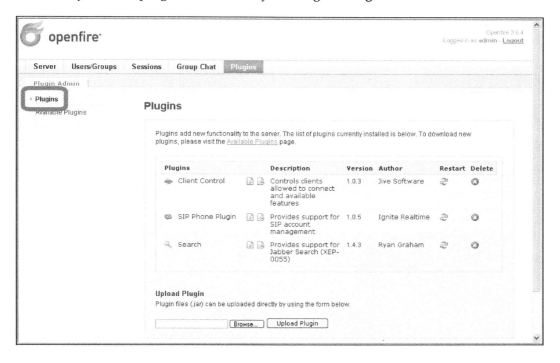

Now, we will create the mappings between Openfire and 3CX by clicking on **Server | Phone | SIP Settings**, as shown in the following screenshot. Set the **SIP server** to the IP address of your 3CX server. The default voicemail number in 3CX is 999, so let's set **Voice Mail Number** to **999**. We can now click on **Update Settings**:

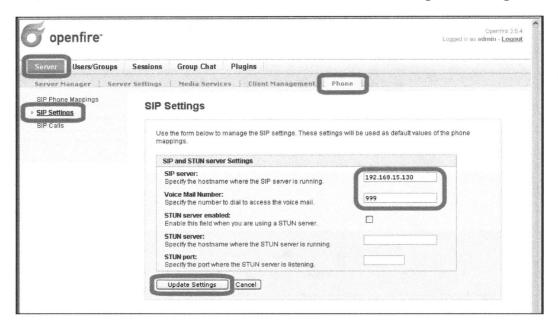

Next, we will map the Openfire users to 3CX extensions by clicking on **SIP Phone Mappings** and then **Add New Mapping**, as shown in the following screenshot. In **XMPP username**, enter the Openfire user and in the **SIP username**, **Authorization Username**, and **Password**, enter the 3CX extension number and passwords, then click **Create**. Do this for each 3CX extension:

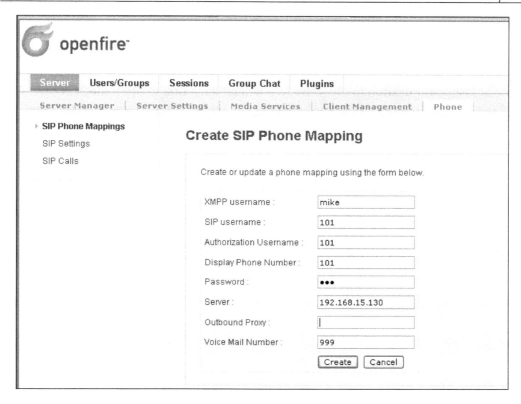

Now, we can click on **SIP Phone Mappings** to see the list of mappings that we created to verify that they are correct:

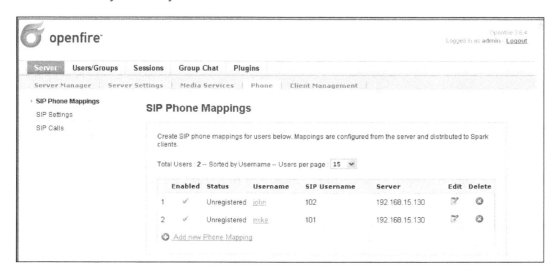

The mappings **Status** will say **Unregistered** even though the setup is correct.

For our last configuration in Openfire, we will create a link to open the 3CX MyPhone page from each Spark IM client. Click on **Server** | **Client Management** | **URL Bookmarks** and then click on **Add URL Bookmark**. Enter your 3CX server IP address in the **URL** field. Check **All Users** and enter the **URL Name** as shown in the following screenshot. Click **Create** and we're finished:

Now, we will move on to installing and configuring the Spark client. First, let's install the Spark IM client on a computer other than the server. We'll log in using one of the users that we set up to make sure that Openfire is configured and working correctly by instant messaging between two IM clients.

Now, let's log on to Spark using one of the users that has a SIP phone mapping. Now let's click on **Spark** | **Plugins** and then install **Phone Client** from **Available Plugins**.

 You may need to restart the Spark IM client for the SIP phone to take effect.

You can now make any call out of 3CX directly from the Spark IM client dial pad:

You can right-click on a contact in your buddy list to call him/her:

You can escalate an instant message conversation to a voice call simply by a click, as shown in the following screenshot:

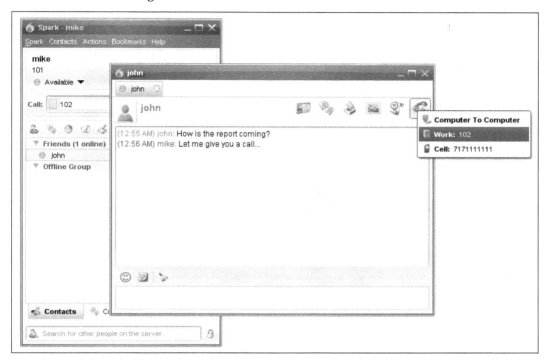

Integrating Legacy PBX

It is not uncommon for a Legacy PBX to need to co-exist, at least for a time, with a new IP PBX. Perhaps management wants to try-before-buying or see the new system working before dismantling their 20-year-old system. As working with Legacy equipment is a reality in everyday life, we will go over integrating 3CX with a Legacy phone system. We will use an ATA device that converts an analog device into a SIP extension, which connects our Legacy PBX to our 3CX Phone System.

At our office, we replaced our four-line Panasonic KX-TG4000B telephone system with 3CX, as shown in the following image. We'll use that as our example, but remember that the process is largely the same even with a much larger system.

We'll use this Legacy four-line Panasonic phone system as our example. As it is a four-line system, we can have up to four concurrent calls between 3CX and this system.

The basic steps are as follows:

1. Configure the 3CX Phone System with a PSTN Gateway.
2. Create extensions in 3CX for each talk path that you need between the Legacy PBX and 3CX.
3. Unplug the PSTN lines from the Legacy PBX line jacks, and plug them into the 3CX PSTN Gateway.
4. Connect the FXS(s) to the Legacy PBX line jack.
5. Configure FXS devices on each of these 3CX extensions.

Following is an overview diagram of how our system will look before and after our integration. We will be putting 3CX between our Legacy PBX and our PSTN or VoIP trunks that connect our systems to the outside world.

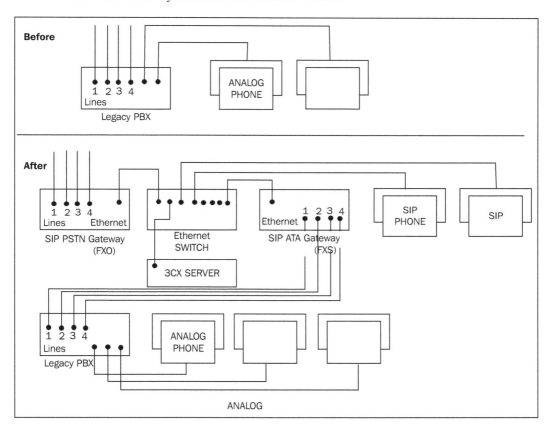

As noted in the diagram, the 3CX Phone System will be inserted between the old phone system and the PSTN lines and when our system is connected and configured; our Legacy PBX will be able to perform these functions:

- Make outgoing calls on PSTN and VoIP trunks
- Receive incoming calls on PSTN and VoIP trunks
- All Legacy PBX extensions can call 3CX extensions

Alright, let's get started. As our 3CX Phone System is already set up and working, we can move right to the next step. Using what we learned in Chapter 3, we'll add extension "200" as the first extension to link our two PBXs. If we name the extension "Legacy Line1", we can see (at least a little bit) where it came from when we call 3CX extensions.

As the 3CX extension "Legacy Line1" will not map to a specific extension on your Legacy PBX, we will probably want to disable voicemail for this 3CX extension. We are using the extension as a way to make a bridge between 3CX and the analog Legacy PBX.

Now, we'll unplug the PSTN line from the Legacy PBX line jack (Line1) and plug it into our 3CX PSTN Gateway. We will plug a telephone cable from the Legacy PBX Line1 jack to the Patton M-ATA telephone jack.

As we just noted, we will use a Patton M-ATA device for this example. It has one FXS port to connect one analog device. GrandStream, AudioCodes, Patton, and other vendors make multiple port FXS devices, if you have a larger implementation.

To start the configuration of the Patton M-ATA, we will plug it into our switch, plug in the power, and verify that the telephone line is plugged into our Panasonic Line1 jack. The Patton M-ATA is set to get an IP address from DHCP. Once it has powered up, we can dial **** and then **100#** using Line1 from any extension on the Panasonic phone system to hear what IP address the Patton M-ATA has acquired. In our case, it received 192.168.15.145, so let's open a browser and type in the following: http://192.168.15.145/. The unit will ask for a password (which is **root** by default); then, click **Authenticate**. You should see the **Home** screen as shown in the following screenshot:

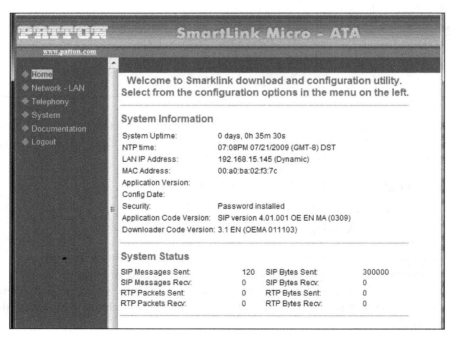

As shown in the following screenshot, click **Telephony | SIP** and in the **SIP Registration Server Address** field, enter the IP address of your 3CX server. Scroll to the bottom of this screen and click **Save SIP Settings**:

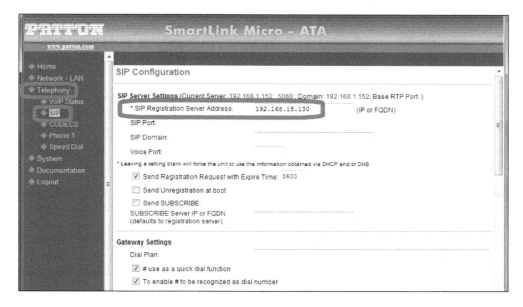

Next, click on **Phone 1** and set **Phone Number**, **User Name**, and **Password** to **200**. Then, click **Save**:

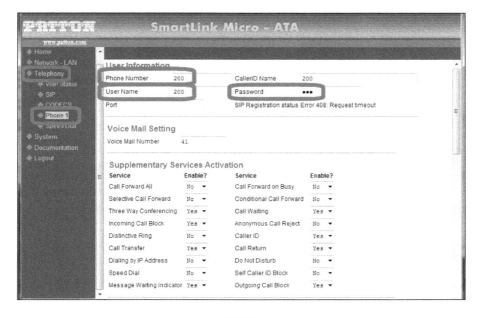

The new settings will not take effect until we reboot this device. So, click on **System | Reload**, making sure **Reset and execute Main Application** is checked, and then click **Reset**. The device will now reboot and, when it is done, your systems should be connected. We'll now test that connectivity.

 For more configuration tips about the Patton M-ATA, you may want to see http://wiki.3cx.com/phone-configuration/vendor-supported/patton-m-ata.

Calling extensions between systems

To test our connection from Legacy PBX to 3CX, we can dial 101 on any of our Panasonic handsets, and extension 101 on the 3CX Phone System should ring.

From 3CX, we can dial extension 200. There are several likely ways that the Legacy PBX will be set up. If you have the luxury of having one line for each Legacy PBX extension, then you can map it so that line X is directed to extension X and you can set up a FXS and extension in 3CX for each extension on the Legacy PBX. This is the cleanest setup. If you have more extensions than lines on the Legacy PBX, then it will either ring all extensions on a certain line and anyone can pick up (not a good way to call a specific extension!), or you will hear a menu and can dial the specific extension you are trying to reach from that menu.

 When calling from a 3CX extension to a Legacy PBX extension when you have more than one line on your Legacy PBX (but not one Legacy PBX line mapped out for each Legacy extension), you may want to create a 3CX Hunt group. Then, add all the 3CX extensions that are acting as bridges to that Hunt group. Make sure the Hunt group **Ring strategy** is selected as **Hunt**. This way, you can remember one number, and it will find a free line that connects to the Legacy PBX.

Outgoing calls over PSTN or VoIP

As 3CX already has outgoing dialing rules set up, calls dialed from the Legacy PBX will work exactly like calls from 3CX. So now, that old PBX can make money, saving VoIP phone calls!

Incoming calls

Incoming calls to the Legacy PBX will experience similar issues to what we discussed earlier in the *Calling extensions between systems* section. If you can have lines mapped to extensions, you can use 3CX digital receptionists to direct calls entirely.

More integration possibilities

We've quickly covered just a few of the many ways to integrate 3CX with other systems. As 3CX uses standard protocols, there are all kinds of ways you can integrate it with other systems. I'll list a few tutorials on the Web that can assist you in integrating with systems that we haven't covered here:

- Exchange 2007 Unified messaging:

  http://wiki.3cx.com/documentation/general/exchange-server-2007

- Connect 3CX directly to other SIP servers:

  http://wiki.3cx.com/documentation/general/sip-dns-configuration

- 3CX and Skype:

  http://wiki.3cx.com/gateway-configuration/vendor-supported/
  skype-to-3cx-part2

Summary

In this chapter, we covered some of the integration possibilities with 3CX and other systems. We covered how we can use Outlook 2007 to initiate calls in 3CX. We tightly integrated 3CX and a full featured instant message system. We also connected 3CX and a Legacy PBX so that we can slowly phase out the older system. We wrapped up by listing some other integration possibilities.

In the next chapter, we will dig a little deeper into the different hardware that interoperates with 3CX such as phone handsets, PSTN and T1 gateways, ATA devices, and more.

9
Hardware

A phone system that has no way of interfacing with the real world is of little use. While you could have a software-based phone system server and softphone, even then you need a headset. The hardware part of the phone systems plays a large part in how users perceive how nicely the phone system works. In this chapter, we will be looking at several hardware components that help make the 3CX Phone System useful such as phone handsets and gateways. In my experience, gateways tend to be the most complicated and time consuming hardware device to set up on your phone system. We will offer some tips to help make that process easier.

Some of the things we'll look at in this chapter are:

- FXO and PRI/T1 gateways to connect our phone system to PSTN lines
- FXS gateways to connect an analog device as an extension to our phone system
- SIP handsets as an extension to the 3CX Phone System
- Firewall configuration

Gateways: The connection to the outside world

Gateways are devices that convert various media into the SIP protocol to work with 3CX or other SIP devices. Gateways allow many devices that were never designed to work together to interoperate smoothly. It also allows you to make use of old equipment and protect your investment in existing equipment.

Looking at the Patton 4114 FXO gateway

The Patton SmartNode 4114 gateway will connect an analog PSTN network and 3CX, allowing you to continue to use your existing phone lines until you move to SIP VoIP trunks.

The Patton SmartNode 4114 gateway interoperates very well with the 3CX Phone System. The unit is housed in plastic with LED indicator lights at the front, along with the **Console** port (which, in a typical installation is not used). On the back side of the unit is the **Power** jack, the Ethernet jack labeled **10/100 ENET**, and four **Voice Ports**, as shown in the following image.

 In our IP PBX consultancy, we've worked with gateways from several different manufacturers: AudioCodes, GrandStream, and Patton. Our experience has been that Patton has delivered the most consistent product and call quality. Also note that AudioCodes is not supported by 3CX. A list of supported gateways can be found at http://www.3cx.com/support/index.html.

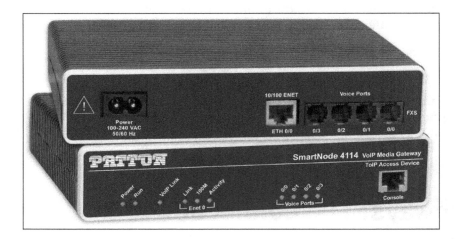

We'd like to quickly note several features that can be tricky for the new user of the Patton gateway. One item is the **VoIP Link** indicator light on the front of the gateway device. Note that this light is lit only when there is a call in progress on the gateway; it does not indicate that the gateway is connected to the 3CX Phone System. To test that, check in the 3CX administrator console.

Also, note that four **Voice Ports** on the gateway are numbered from 0 to 3. On the back of the unit, note that the **Voice Ports** start at port 3 (the fourth port) on the left. This will be important if you are not using all the ports.

Configuring the Patton 4114 FXO gateway

We noted earlier that configuring gateways is possibly the most complex part of setting up your IP PBX. Of all the software IP PBXs I've worked with, 3CX makes this process the most simple by making a configuration file for many popular gateways, including this gateway. So the configuration consists of making sure that the gateway has been updated to the correct firmware version and importing the configuration file generated by 3CX.

Configuring the gateway in 3CX

The first step in getting your gateway set up is in the 3CX Phone System. Let's click on **Add PSTN Gateway**. Now, we can type in the **Name** we want to give this gateway (for example, **PSTN_Gateway**). We can select our gateway **Brand** and **Model** from the drop-down lists and then click **Next**, as shown in the following screenshot:

The firmware version is very important because the configuration file is different for different firmware versions. Note the firmware now, so that we have it for reference later. You can find the latest recommended firmware at `http://www.3cx.com/voip-gateways/patton-smartnode.html`.

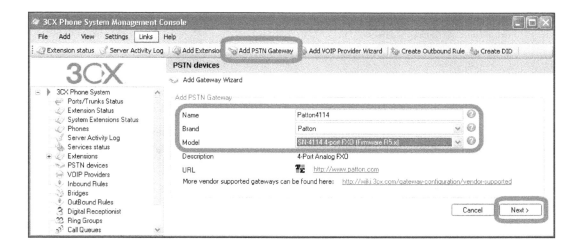

The next screen will let you edit the way the gateway works. The most important field here is the **Country**. Select yours and click **Next**:

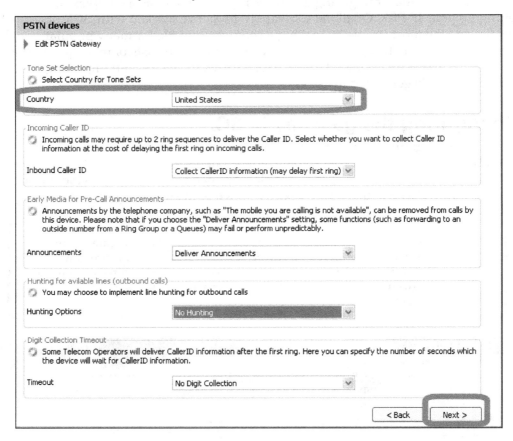

On both the **Create Ports** screen and **Outbound Call Rules** screen, click **Next**. You will get to the following screenshot once you do that. Enter the IP of your gateway and press **Next**:

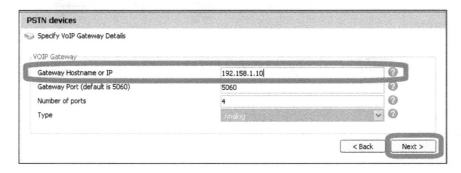

When you are finished configuring the gateway, the **Gateway Created** screen will be displayed. This screen will show a summary of the gateway that was just created. At the bottom of this window will be the **Generate config file** button which will create a configuration file that we will use when configuring our gateway. Now we are done in 3CX.

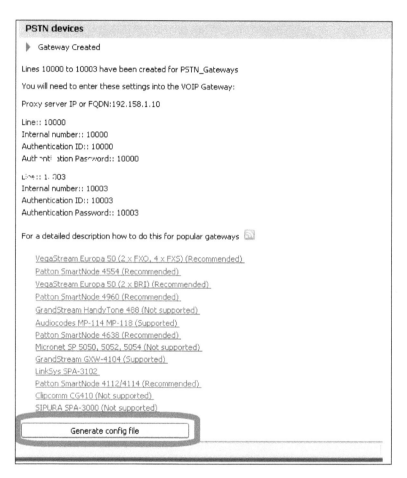

Getting the Patton gateway on your network

First of all, plug the gateway into your switch and then power up the gateway.

On the CD-ROM that came with the Patton 4114, run
`E:\TOOLS_WI\SNDiscovery.exe` (where `E:\` is your CD-ROM drive).

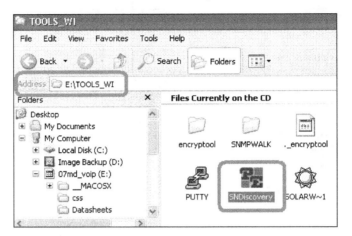

The Patton 4114 is set to receive an IP address from DHCP by default, so you will need to have a DHCP server running on your network. The `SNdiscovery.exe` tool will find what IP address DHCP gave your Patton 4114 gateway, as shown in the following screenshot:

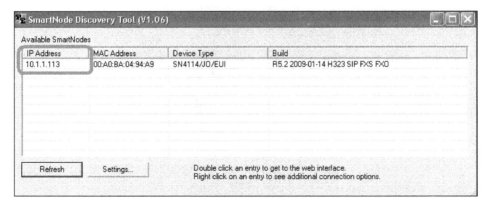

If there is more than one Patton SmartNode gateway on your network, you will want to make sure the **MAC Address** matches the one we are working with. Next, we will log in to the Patton 4114 gateway via a web browser and configure it with a configuration file.

If you assign the Patton gateway an IP address in the configuration file (which you will want to do), it will no longer get an IP address from DHCP and will change after you configure the Patton gateway. Keep that in mind.

Making sure the Patton gateway has correct firmware

The Patton gateway needs to be using the firmware designated in 3CX to work with the configuration file we just created. If your gateway is newer or older, we need to load the correct firmware version.

To get the correct firmware version, visit `http://upgrades.patton.com`.

 You will need a Patton support username and password to get this file.

The Patton gateway firmware file you download will be a ZIP file. Do not extract it as the Patton gateway firmware update expects it to be a ZIP file only. Save the file and log in to the Patton gateway.

To log in to the Patton gateway, just open your Internet browser and type in the IP address. The default username is **administrator** and the password can be left blank.

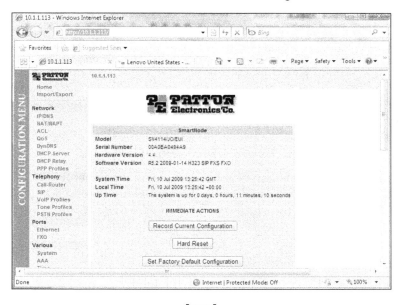

If you log in successfully, you will see the **Home** page of the Patton gateway as shown in the previous screenshot.

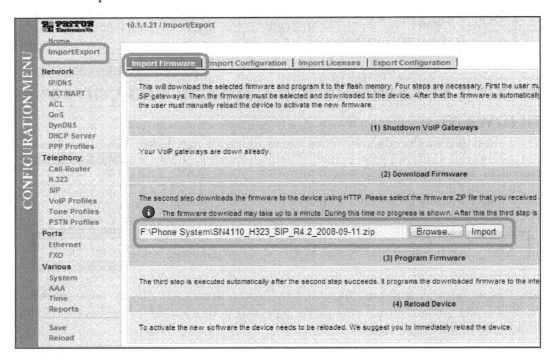

Click on **Import/Export | Import Firmware**.

 If this is not a new gateway, there may be SIP gateways in use and you will need to click on the **Shutdown SIP Gateway** button.

Now, **Browse** to the firmware (ZIP file) that you downloaded and click **Import**. The gateway will now load the firmware and will give you the progress as it does that. *Do not power off the gateway during the update*. When it is complete, you can click **Reload** to restart the gateway and load the configuration.

When the gateway reboots, open a new browser, log in again, and check the firmware version on the **Home** page to make sure that the firmware loaded successfully.

Configuring the Patton gateway

Let's import the configuration file that we created earlier by clicking **Import/Export** on the Patton web interface. Then, click **Import Configuration**:

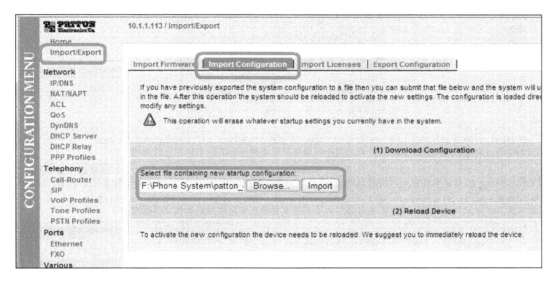

Click **Browse** and select the text configuration file we made earlier. Click **Import** as shown in the previous screenshot.

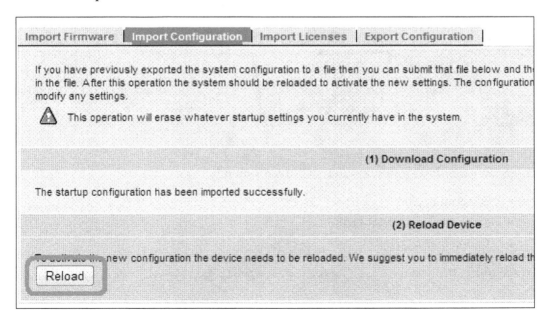

Now, you can click **Reload** to load this configuration into the gateway. You will be asked to verify that you want to **Reload**, when you do that, the gateway will reboot. The gateway will take just a minute or two to reboot;

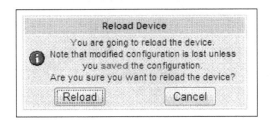

Log in to the Patton 4114 with the new IP address just to make sure your configuration took effect.

Following is the image of a Patton 4114 unit. Note that the **VoIP Link** LED is lit only when a call is in session on the unit and not when it is merely registered to 3CX:

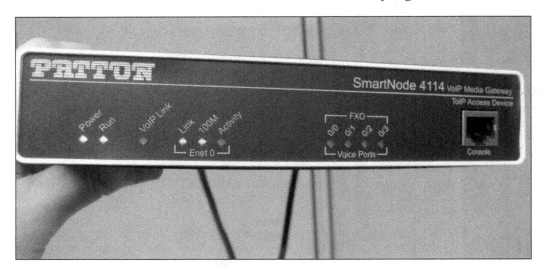

There are other FXO gateways that 3CX recommends:

- Sangoma A200
- GrandStream GXW-410X
- Linksys SPA-3102

To see more, browse to `http://www.3cx.com/voip-gateways/index.html`.

Looking at the Patton 4960 T1 gateway

The Patton 4960 allows you to connect 3CX to T1 or PRI depending on the model you get. It can have one to four ports. Following is the image of a Patton 4960 unit with four PRI ports and two Ethernet ports:

Configuring the Patton 4960

The Patton 4960 configures exactly like Patton 4114. The user interface and steps to import the 3CX configuration file are identical, so we can use the steps mentioned previously in the *Configuring the Patton 4114 FXO gateway* section to import the 3CX configuration file.

The one you have to watch out for is related to the Ethernet ports on two port models. When the Patton gateway ships, new port 0/0 will be a DHCP client and port 0/1 will be a DHCP server, so you will want to plug into port 0/0. The tricky thing is after the gateway has been configured, data will be routed to port 0/1, so you will need to switch this after the unit has been configured by the 3CX configuration file.

Some other 3CX supported T1/PRI cards include:

- Patton SmartNode 4554/4960/4638
- Sangoma A101 T1/E1/J1
- BeroNet BeroFIX 400/1600/6400

The latest supported gateways can be found on the following page: http://www.3cx.com/support/index.html.

ATA connects your analog devices to your PBX

An ATA gateway allows us to connect any analog phone device to our 3CX Phone System. In essence, it converts from an analog phone signal to SIP, letting us use, for example, any standard analog phone handset with our 3CX IP PBX! You could use your long range portable phone, a bunch of handsets that seem a waste to throw away, a fax machine, or any other analog phone device. Once again this type of gateway allows us to use equipment that we've already invested in.

 Yes, you can connect an analog fax machine to your phone system but the ATA you select must support T.38 protocol. Also, your FXO must support T.38 and even then it is tricky to get it all working. I'd recommend keeping the fax machine outside your PBX (if possible) or using the 3CX fax server.

With an ATA gateway no phone is too old to connect to your 3CX IP PBX. An ATA will allow you to convert any phone to a SIP phone.

Looking at the Patton M-ATA

The Patton M-ATA is quite compact and fairly simple. On one side is the power and telephone jack that connects the M-ATA to your analog phone or device. The other side has an Ethernet jack to connect the gateway to your network. There is also a small row of LEDs' to show the status of the device. The following image is of a Patton M-ATA:

Configuring the Patton M-ATA

To start configuring the Patton M-ATA to use an analog phone as an extension, we will plug it into the switch, plug in the power, and verify that the telephone line is plugged into our analog phone. The Patton M-ATA is set to get an IP address from DHCP. Once it has powered up, we can dial **** and then **100#** using the analog phone to hear what IP address the Patton M-ATA has acquired. In our case, it received 192.168.15.145, so let's open a browser and type in http://192.168.15.145/. The unit will ask for a password (which is **root** by default), then click **Authenticate**; you should see the home screen.

As shown in the following screenshot, click **Telephony | SIP**, and then enter the IP address of your 3CX server in the **SIP Registration Server Address** field. Scroll to the bottom of this screen and click **Save SIP Settings**:

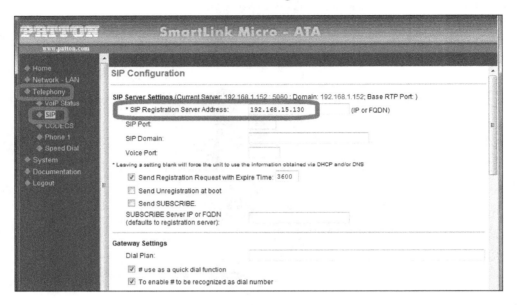

Next, click on **Phone 1**. Set **Phone Number**, **User Name**, and **Password** to **200**, and then click **Save**, as shown in the following screenshot:

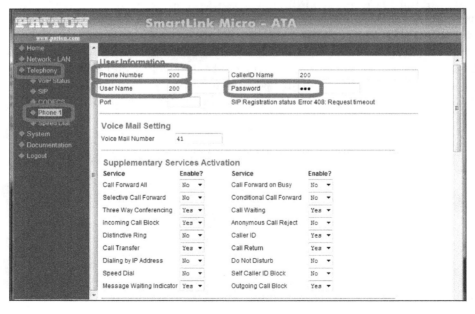

The new settings will not take effect until we reboot this device, so click on **System | Reload**, making sure **Reset and execute Main Application** is checked, then click **Reset**. The device will now reboot and, when it is done, you are done configuring the device and we can create an extension in 3CX using the steps we noted in the *Basic extension setup in the administrator console* section of Chapter 3.

 For more configuration tips, you may want to see `http://wiki.3cx.com/phone-configuration/vendor-supported/patton-m-ata`.

Now that we realize that we can connect cheap analog phones to our 3CX Phone System, we might be eager to do this instead of buying true SIP phones. Keep in mind, however, that a low-end SIP phone is usually not that much more expensive than an ATA. Plus, an analog phone will typically not have nice buttons for putting calls on hold, transferring calls, or showing BLF indication. Many features like transfer, hold, and so on are possible by using star codes in 3CX but will require more button presses and are not as handy.

 For more ATA gateways that work with 3CX browse to `http://www.3cx.com/voip-gateways/index.html`.

SIP phone handset

As we noted earlier, SIP phone handsets are mainly used for interaction with the 3CX Phone System. We won't be able to look at every SIP phone, but we will look at one that is fairly representative of SIP handsets.

Looking at the Snom 360

The Snom 360 is very representative of popular SIP handsets and comes in a traditional form factor. Most functions are available on a single key press. The phone includes 12 BLF indicators and buttons that can be configured to monitor the status of other extensions. The phone also includes a **Message Waiting Indicator** (**MWI**) lamp to indicate voicemail and missed calls.

Now, we'll take a look at the back of the Snom 360, as shown in the next screenshot. The first jack moving left to right is the expansion module plug. This allows us to add 44 more BLF indicators and buttons. These expansion modules can be daisy chained to add up to three. The next jack is the power jack. This phone can be powered by a power cord or via Power over Ethernet.

 Don't over look getting a Power over Ethernet enabled switch. One very nice ability it gives you is that you can plug your switch into a battery backup UPS and have all your phones stay online if the power goes off! Just like in the good old days!

The third jack is where you plug your phone into your network. The fourth jack is an Ethernet hub and will allow you to plug your PC into it, requiring only one Ethernet cable to energize both your phone and PC.

 I would recommend getting two Ethernet cables for your desks. Most SIP phones provide a mere 100MBps network connection. Another negative aspect is that if you reboot your phone, for administrative purposes or whatever, your PC will be disconnected.

Moving right is the jack to plug in a Snom headset and the last jack is for the phone handset.

Following is an image of the Snom expansion module which is great for receptionists. These modules can be daisy chained up to three for quite a bit of expansion:

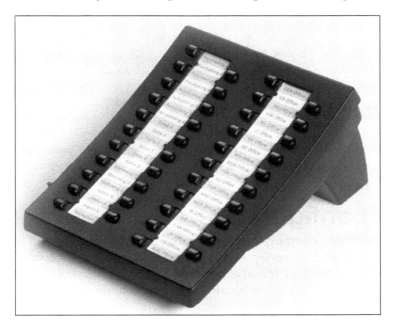

The Snom 360 form factor is a very good phone to replace an existing phone system for users who want to have single button access to features and the ability to monitor other extensions.

There is also another phone format that has fewer BLF lights and depends more on soft keys. The Linksys SPA941, Snom 820, and many other modern phones fall into this category. Note as shown in the following image, the **Linksys IP Phone SPA941** depends on soft keys for the transfer button and there are only four possible built-in BLF indicators:

Configuring the Snom 360

Configuring the Snom 360 and many other handsets is done via a web interface. We covered configuring the Snom 360 in detail in the *Connecting a Snom 360 phone* section of Chapter 3, so we won't go over that again in this chapter.

Router configuration

The biggest challenge is getting the gateway and the router or firewall in your phone system network to work correctly. For your first 3CX install, I'd recommend using a simple firewall to make your initial experience less tricky. We will be using a simple but hardy Linksys WRT54G as our device in the demonstration.

Looking at the Linksys WRT54G

The Linksys WRT54G is very simple to set up and has a simple built-in DHCP server that will be handy for assigning phones and other devices' IP addresses. Set up the Linksys just like you would for a normal network.

The Linksys WAP54G is a very simple router that is great for getting started with 3CX. Amazingly, more robust firewalls will be more complex to get going with 3CX.

A very common mistake made related to routers is having two routers in a row. A common reason why this happens is that the modem from the ISP is also a router and the network administrator puts a small Wi-Fi firewall in too, effectively making all SIP traffic pass through two routers. I would highly recommend avoiding this scenario. It is hard enough to get VoIP traffic to pass through one router!

To configure the Linksys, log in to the unit and click on **Applications & Gaming | Port Range Forwarding**. Set up the three port forwarding rules, as shown in the following screenshot with the IP address of your 3CX server:

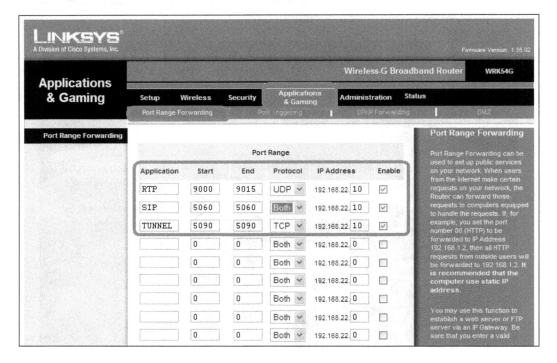

Now to verify that you have forwarded all the ports correctly, log in to 3CX and run the firewall checker.

There are many other firewalls that can work with 3CX. Only requirement is that the firewall supports **full coned NAT** to properly handle VoIP traffic through the router.

Summary

We've taken a quick tour of some commonly used 3CX hardware such as FXO gateways, to connect PSTN phone lines to your phone system. In addition, we learned how to use a PRI gateway and an ATA gateway to connect any analog device as an extension, and we looked at some of the features of a SIP handset. We also noted that 3CX does a great job of creating provisioning files for these devices. Finally, we took a quick look at setting up a router to work with 3CX.

10
Maintenance and Troubleshooting

Maintenance and troubleshooting—two very good topics! You will of course, have to maintain your system; it's not going to be a simple, install and forget it for years kind of system. Why not? Well because it's running on Windows, and that of course will change. Right now, Windows 7 is out. Will it run 3CX better than XP, Vista, or Server 2003/2008? Not sure yet. You will have Windows updates for security, firewall issues to deal with, especially if you are using an ITSP. But what if you start off with analog lines and want to change to VoIP? What if new phones come out that are better? Think of video conferencing with color touchscreens built-in! Remote locations, the obvious add, move, or remove users, maintenance, or even a new version of 3CX will be out with new features that you might want to take advantage of, for whatever reason. 3CX, or really any software-based VoIP PBX system will be updated with new features.

Don't forget hardware failure. Hard drives are bigger and faster than ever, but they still crash. They all will, it's just a matter of time. So, you will want to do backups, you will want to have some redundancy. You will need some help figuring out why something doesn't work like you expected.

This chapter covers all of that. First, we will start with the unpleasant topic of crashing.

Disaster recovery

Let's start with backing up your configuration. Want to do a small upgrade? Backup your configuration. Want to do a configuration change? Backup your configuration first, just in case you need to revert to the previous one.

There are a couple of ways you can do backups. One way is a full system backup, which will cover Windows, 3CX, your call logs, and everything you'd need to recreate your system. How do you do this? There are several ways and they all have their pros and cons.

One such way is **Windows backup software** — Yes, Windows has a decent backup software built-in. No, it's not the best, or easiest, but it's free. Starting with Vista and Server 2008, it's much better, as long as you want it to be backed up to locally attached hard drives. It works well, but it's not my first choice.

There are lots of third party backup software products out there. They are generally better than the Windows one, however they cost money, sometimes quite a bit, if you want to backup open files. What are **Open files**? These are files that are in use by either the system or a user and cannot be backed up easily. The open file agents are needed to backup these types of files.

I'm starting to use online backups more and more. There are companies out there that can backup your data, even open files, and store it securely on a data center. As long as you have a decent Internet connection (think upload speed here), this will work well. All your data files will be encrypted and compressed off-site. They are also automated, so you don't have to do anything once the schedule is set. Only problem is they usually only back up the data, which is the important part. A full restore, however, can take some time.

So, that's a brief overview of some system backups. What about just backing up 3CX? Of course, you will want to do that from time to time. This is a configuration backup and also recommended before you do any changes or software upgrades.

The only issue I have with the backup function is that it's done from the 3CX machine itself, not from the web interface, so it's as convenient.

To do a local backup of the 3CX configuration, you will need to be on the computer having 3CX installed. From there, open the 3CX program group and select **Backup and Restore Tool**:

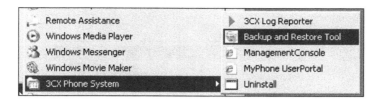

On the next screen, as shown in the following screenshot, you will see a couple of options in the backup tool. Select all of the items you want to backup. I like to do everything if time permits. However, if you have gigs of voicemail and a year's worth of call logs, it will take quite a while and use a lot of space.

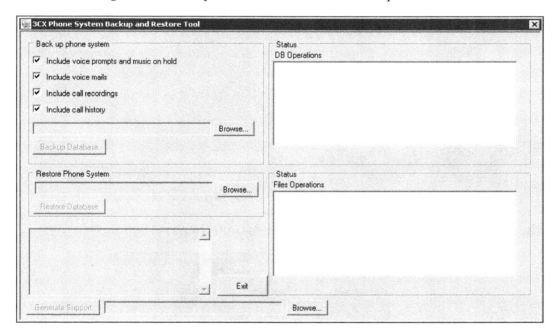

Now, click our familiar **Browse** button in the **Back up phone system** section to find the location where we want to store the file and then name the file. This can be any drive that you have access to and permission to use. Network drives, USB hard drives, or flash drives (whatever you like) as long as Windows can see it, and you have write permissions on it:

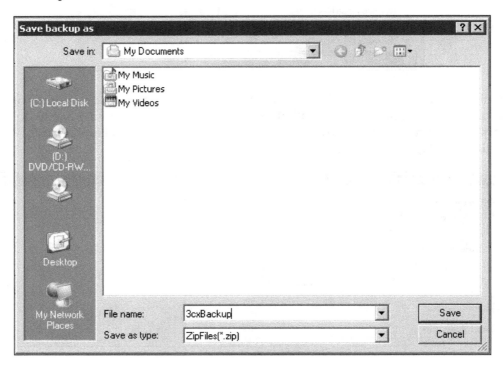

Now go ahead and click **Save**. Then click the **Backup Database** button:

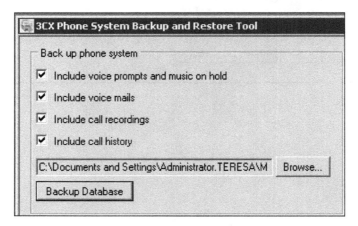

This will start the backup process. You will see the status of the backup process on the right-hand side panes. When it is done, you will see a **Operation Successful** message:

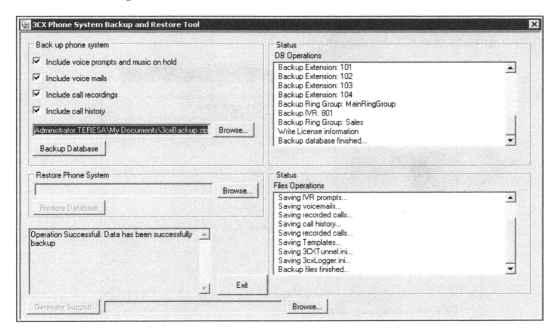

To restore your configuration to a previously created backup, you will need to click **Browse** under the **Restore Phone System** section and locate your backup file, then click **Restore Database**. This will OVERWRITE all your configurations you have made since the last backup.

You also have a **Generate Support** option, click the **Browse** button to set the name and location and click **Generate Support**. This will save a configuration file and the log files needed for 3CX support to see what is going on with the system. It is a great time saver over trying to e-mail/phone error messages:

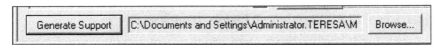

One of the drawbacks of this utility is that it cannot be scheduled. So, 3CX has a command-line version of this tool, which can be used to schedule tasks to automate this process. Start by launching **Control Panel | Scheduled Tasks | Add Scheduled Task**. This will start a wizard, as shown in the following screenshot. Click **Next**, then **Browse** to find the 3cxbackup.exe program. The default location is C:\program files\3CX PhoneSystem\Bin.

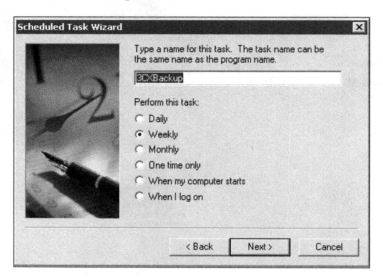

After that, set your days and times on the next few screens. Don't forget to use the correct username and password for that PC so that the program has permission to run the backup.

Now, open the advanced settings for this task and add the appropriate options at the end after the 3cxbackup.exe:

- /hidden will run the backup and then close it when it is done
- Backup or Restore depending on what you want to do
- {filepath} full file path of where you want the backup to go
- /history will backup the call history database
- /prompts will backup the system prompts (important if you have any custom ones)
- /recordings will backup all the call recordings

The following screenshot shows what the backup will look like. You can see in the **Run** field, I added "/hidden backup c:\3CXbackup.zip /prompts /history" (after 3cxbackup.exe):

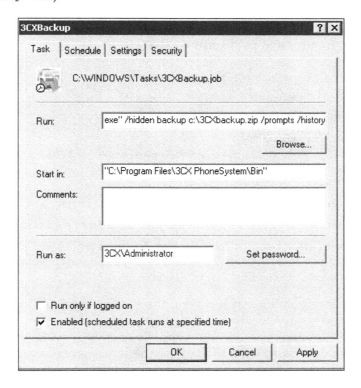

This scheduled task should now run at the day(s) and time you specified, just check it and make sure it works as expected.

With any backup type that you use such as software, online, scheduled, and so on, try to restore a file once in a while to make sure it works. Nothing is worse than thinking that you have secure backups and then finding them all blank, or realizing your drive was bad and nothing was backed up.

Another newer backup option is **imaging software**. There are third party applications available that can take a snapshot of the entire hard drive and save it somewhere. This snapshot can be used to do a complete restore of a failed PC usually in 10-20 minutes. Do a web search for such imaging software. They range in price from free an open source software to thousands of dollars. Following is a small list based on open source and paid software that I've used:

- Open source:
  - **FOG**: http://www.fogproject.org
  - **Clonezilla**: http://www.clonezilla.org
  - **DriveImage XML**: http://www.runtime.org/driveimage-xml.htm

- Paid software:
  - **Symantec Ghost**: http://www.symantec.com
  - **Acronis Backup and Recovery**: http://www.acronis.com
  - **Symantec Backup Exec System Recovery**: http://www.symantec.com/business/backup-exec-system-recovery-server-edition

Other disaster recovery options include Windows clustering (Windows Enterprise Server is needed for this) and virtualization. Some people have good luck with virtualization but others complain of choppy audio and lags in software. It all depends on what kind of use the 3CX system has and the hardware running everything.

Trunk backup

In previous chapters, we discussed trunks and dial plans. How do you backup those if one goes down? You have to set up several trunks and then, in your dial plan, set the other trunks as 2nd or 3rd options:

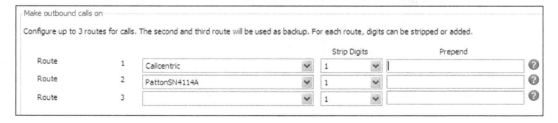

Just make sure your **Strip Digits** and **Prepend** rules apply correctly to those lines.

That covers outbound backup, but what about inbound? This is the tricky part. You will need to talk to your PSTN or ITSP vendor and see what options they offer. Some have roll-over options, where the call will go to another number in their system.

Others have a "disaster recovery" option, where they will forward the incoming call to any phone number you want. Take advantage of these options as your lines will go down someday.

Firewalls

If you are using PSTN lines, feel free to skip this topic as you don't have to worry about any firewalls, NAT, or ports configurations.

Unless you stick your non-firewalled PBX system on the public network with a public IP, you will need to adjust your firewall. The firewall is usually your first line of defense in isolating your computers from the outside world. These can be hardware (Linksys/Cisco, Netgear, D-Link, among others) connected to your Internet modem, or software (Windows Firewall, or some other third-party software) firewalls that you will need to adjust to allow the SIP traffic into your private network.

Most people use **Network Address Translation (NAT)** in a small office/home environment. This makes it easy to set up, and you don't have to buy a bunch of static IPs' for various server uses (e-mail server, web server, PBC, and so on).

You will need to forward the following ports to your PBX:

Port	Description
UDP 5060	This is the standard SIP port used to send and receive calls.
TCP 5090	This is for 3CX remote extension tunneling.
UDP 9000-xxxx	These ports are needed for the RTP packets for the actual call. You will need two ports for each supported call. So if you need ten calls going in/out of your system, you will need to open 9000-9019 (20 ports).

That's it for the outside world. For the internal calls, you will also need to either disable the firewall or open these ports:

Port	Description
UDP 5060, 5480, 5482, 5483, 5485, 5487	This is for the 3CX system
TCP 5090	For the remote 3CX tunneled extensions
UDP 7000-7500	This range is for the internal calls
UDP 9000-xxxx	The same as what is needed for the outside world you set up above

These are the default ports in 3CX. You can change them if needed, but it's not really a good idea. Troubleshooting is harder and if you don't use 5060, your VoIP provider might not support you.

Your other option is to use a public IP and have your PBX server out on the Internet. This is fine only if you have a firewall in place that blocks all the other ports. Ideally, this is a better setup since NAT can cause problems. Not all routers support NAT and SIP. Some can do the port forwarding just fine, but some cannot. This can cause loss of audio—inbound and/or outbound—also called one-way audio.

Newer routers, or older ones with newer firmware, support **SIP-Application Layer Gateway (SIP-ALG)** and can NAT our 3CX system just fine. I'd recommend one built for VoIP traffic that can prioritize those particular packets. Check the manufacturer's website to see the latest updates and support for SIP-ALG. As SIP becomes more popular, even older models are getting firmware upgrades to support SIP-ALG.

Using logs to troubleshoot your phone system

The 3CX **Server Activity Log** will provide detailed information about exactly what is happening in our 3CX Phone System and can be a great help in tracking down elusive issues. We can take a look at the **Activity Log** by clicking on the **Server Activity Log** link, as shown in the following screenshot. We can also navigate by clicking **View | Server Activity Log** using the drop-down menus.

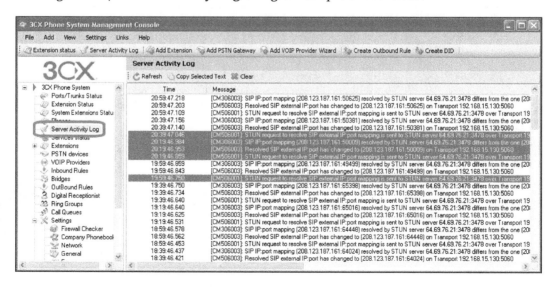

The **Server Activity Log** is fairly self explanatory. Note that the new events will not show up in the log until you press **Refresh**. Another nice feature is the ability to copy individual log entries using the standard Windows *Ctrl + click* convention to select non-consecutive log entries, as shown previously which can then easily be saved to a text file.

As verbose logging is CPU intensive, 3CX is set to **Medium** logging level, by default. If you want more detailed logging, click on **Settings | Advanced** and then on the **Advanced** tab, change the **Logging level**, as shown in the following screenshot. For this to take effect, you will need to restart the 3CX Phone System service using the **Services status** window.

Changes to logging level only take affect after the 3CX Phone System service is restarted.

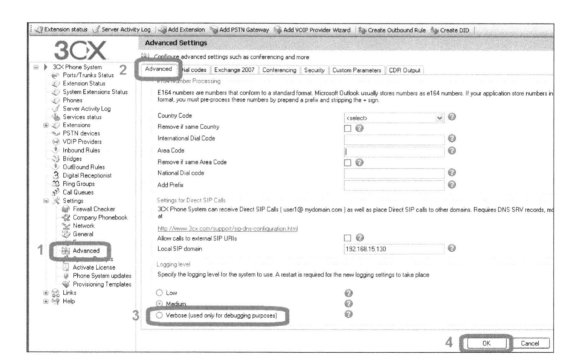

The **Verbose** logging level will take high levels of CPU so enable it only during your troubleshooting. Don't forget to set this back to medium or lower when you are done troubleshooting.

3CX services: They all need to run

3CX uses "services" to run the software. A computer service is a program that runs without anyone launching it. Internet Explorer and MS Word are NOT programs that run as a "service", as you need to log on to the computer and double-click the program to launch them for use.

These services will run all the time (as long as the computer is on, of course) without anyone touching them. 3CX relies on 12 such services to operate its every feature.

In the left-hand window, click **Services status**, and you will see the following screenshot:

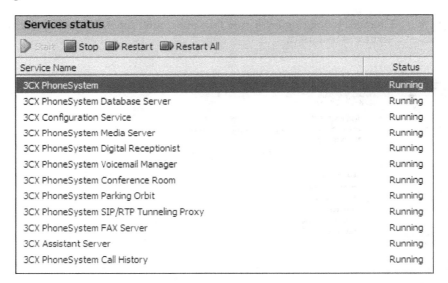

This screen shows you all the services that are installed and if they are running or not. If you cannot make a conference call, make sure this screen shows you that the **Conference Room** service is **Running**. On the top, you have controls to **Start**, **Stop**, **Restart** (stop the service and then automatically start it), or **Restart All**.

If you are experiencing trouble with 3CX, restarting the services can usually fix the problem. It's the same as rebooting the computer but faster, and it only restarts the 3CX programs.

Monitoring

Another useful monitoring tool is Performance Monitor. If you open Performance and add a counter, you will see **3CX PerfMon** listed on the drop-down menu. There you will see some great monitoring options. The number of calls completed (active, successful, or failed), and totals of other items are all worth looking at on a regular basis. Turning these on will take up some CPU time, so enable only the ones you need or check it often to keep track of how your system is doing.

> Performance Monitoring unfortunately does not appear to be a high priority and has stopped working in some versions. Before you spend a lot of time trying to read these, make sure they work in the version you have.
>
> See http://www.3cx.com/forums/performance-monitors-11913.html#p62702 for more information.

When you need support

The best place for free support is the Web.

Wiki.3cx.com is where you will find almost all of the support that you need. Wiki.3cx.com/documentation has the documentation and FAQ available, and the forums at www.3cx.com/forums are a great place with thousands of posts from all over the world. The search tool in the forums is your friend. Still if you can't find the help you need, post a question. You can usually get a response from someone within a couple of hours.

YouTube is also a place to find videos from 3CX and Landis Computer. Videos are updated often, so keep checking if you don't find what you are looking for.

Blogs we find helpful:

- **3CX blog**: http://www.3cx.com//blog
- **Worksighted blog**: http://3cxblog.worksighted.com
- **Deerfield blog**: http://www.3cxblog.com
- **Matt Landis' blog**: http://www.windowspbx.blogspot.com

Classroom 3CX training is also available for those who like a more structured class environment with live instructors and practice equipment at http://training.3cx.com. This is probably the best way to learn, besides this book of course, if you have no experience in 3CX or VoIP/PBX systems.

When your system is down and you want quick, knowledgeable support, you can always pay for it. You can get support from a 3CX partner such as Landis Computer at http://www.landiscomputer.com, or TechNet Computing at http://www.technetcomputing.com or directly from 3CX. They will give you support options via remote access, e-mail, chat, and phone. Prices vary depending on the service you'd like, but it might be worth paying for, if you get stuck. I recommend this option for a business. 3CX recommends that you use a partner; usually the one you buy 3CX from is the best place to start. Remote support can do wonders when you're in a jam. Even if you want a remote installation, your reseller can help you with that as well.

Summary

This chapter covered some important topics such as backing up, disaster recovery options, and how to plan for failed phone lines.

We also covered support and monitoring options. Computers will crash, but knowing how to avoid it and monitor it can lead to better uptimes and a higher return on investments.

At some point, you will need support. You might have a new model router or phone that you are having issues with, or you may want to ask someone about recommendations for hardware. The forum, Wiki, FAQ, and resellers/partners are all available for any situation.

Index

[PACKT] PUBLISHING

Thank you for buying
The 3CX IP PBX Tutorial

Packt Open Source Project Royalties

When we sell a book written on an Open Source project, we pay a royalty directly to that project. Therefore by purchasing The 3CX IP PBX Tutorial, Packt will have given some of the money received to the 3CX project.

In the long term, we see ourselves and you—customers and readers of our books—as part of the Open Source ecosystem, providing sustainable revenue for the projects we publish on. Our aim at Packt is to establish publishing royalties as an essential part of the service and support a business model that sustains Open Source.

If you're working with an Open Source project that you would like us to publish on, and subsequently pay royalties to, please get in touch with us.

Writing for Packt

We welcome all inquiries from people who are interested in authoring. Book proposals should be sent to author@packtpub.com. If your book idea is still at an early stage and you would like to discuss it first before writing a formal book proposal, contact us; one of our commissioning editors will get in touch with you.

We're not just looking for published authors; if you have strong technical skills but no writing experience, our experienced editors can help you develop a writing career, or simply get some additional reward for your expertise.

About Packt Publishing

Packt, pronounced 'packed', published its first book "Mastering phpMyAdmin for Effective MySQL Management" in April 2004 and subsequently continued to specialize in publishing highly focused books on specific technologies and solutions.

Our books and publications share the experiences of your fellow IT professionals in adapting and customizing today's systems, applications, and frameworks. Our solution-based books give you the knowledge and power to customize the software and technologies you're using to get the job done. Packt books are more specific and less general than the IT books you have seen in the past. Our unique business model allows us to bring you more focused information, giving you more of what you need to know, and less of what you don't.

Packt is a modern, yet unique publishing company, which focuses on producing quality, cutting-edge books for communities of developers, administrators, and newbies alike. For more information, please visit our website: www.PacktPub.com.

FreePBX 2.5 Powerful Telephony Solutions

ISBN: 978-1-847194-72-5   Paperback: 292 pages

Configure, deploy, and maintain an enterprise-class VoIP PBX

1. Fully configure an Asterisk PBX without editing the individual text-based configuration files

2. Add enterprise-class features such as voicemail, least-cost routing, and digital receptionists to your system

3. Secure your PBX against intrusion by managing MySQL passwords, FreePBX administrative accounts, account permissions, and unauthenticated calls

4. Packed with step-by-step instructions, examples, screenshots, and diagrams

Building Telephony Systems with OpenSER

ISBN: 978-1-847193-73-5   Paperback: 324 pages

A step-by-step guide to building a high performance Telephony System

1. Install, configure, and troubleshoot OpenSER

2. Use OpenSER to build next generation VOIP networks from scratch

3. Learn and understand SIP Protocol and its functionality

4. Integrate MySQL with OpenSER

5. Integrate OpenSER & Asterisk

Please check **www.PacktPub.com** for information on our titles

PUBLISHING

Zenoss Core Network and System Monitoring

ISBN: 978-1-847194-28-2   Paperback: 280 pages

A step-by-step guide to configuring, using, and adapting this free Open Source network monitoring system - with a Foreword by Mark R. Hinkle, VP of Community Zenoss Inc.

1. Discover, manage, and monitor IT resources

2. Build custom event processing and alerting rules

3. Configure Zenoss Core via an easy to use web interface

4. Drag and drop dashboard portlets with Google Maps integration

Cacti 0.8 Network Monitoring

ISBN: 978-1-847195-96-8   Paperback: 132 pages

Monitor your network with ease!

1. Install and setup Cacti to monitor your network and assign permissions to this setup in no time at all

2. Create, edit, test, and host a graph template to customize your output graph

3. Create new data input methods, SNMP, and Script XML data query

4. Full of screenshots and step-by-step instructions to monitor your network with Cacti

Please check **www.PacktPub.com** for information on our titles

www.ingramcontent.com/pod-product-compliance
Lightning Source LLC
Chambersburg PA
CBHW060551060326
40690CB00017B/3676